4th International Conference on Building Materials and Materials Engineering (ICBMM 2020)

Held online due to COVID-19

IOP Conference Series: Materials Science and Engineering
Volume 1054

Barcelona, Spain
24 - 26 September 2020

ISBN: 978-1-7138-3337-6
ISSN: 1757-8981

Printed from e-media with permission by:

Curran Associates, Inc.
57 Morehouse Lane
Red Hook, NY 12571

Some format issues inherent in the e-media version may also appear in this print version.

Institute of Physics
Dirac House, Temple Back
Bristol BS1 6BE UK

Phone: 44 1 17 929 7481
Fax: 44 1 17 920 0979

techtracking@iop.org

Additional copies of this publication are available from:

Curran Associates, Inc.
57 Morehouse Lane
Red Hook, NY 12571 USA
Phone: 845-758-0400
Fax: 845-758-2633
Email: curran@proceedings.com
Web: www.proceedings.com

TABLE OF CONTENTS

BUILDING MATERIALS AND TECHNOLOGY

STRUCTURAL AND HYDRAULIC ENGINEERING

IOP Conf. Series: Materials Science and Engineering 1054 (2021) 011001 doi:10.1088/1757-899X/1054/1/011001

Preface

The 4th International Conference on Building Materials and Materials Engineering (ICBMM 2020) was planned to be held in Barcelona, Spain. However, due to COVID-19 pandemic, the conference was held virtually in web streaming on September 24-26, 2020. The whole conference was held on-line in ZOOM with a success. ICBMM 2020 conference was organized by the Chemistry and Materials Society (CMS), a branch of Hong Kong Chemical, Biological & Environmental Engineering Society (HKCBEES). Considering the pandemic situation, the organization opted not to postpone or cancel the conference and has created a virtual environment where the attendees presented the papers and participated in the sessions, and where activities that have been organized online can be accessed.

ICBMM 2020 had two keynote speakers presenting the latest developments and trends in their respective fields, Prof. Carlos Chastre from NOVA University of Lisbon, and Prof. Paulo Mendonça, from University of Minho, both from Portugal. Each keynote speech lasted for 50 minutes including 5 minutes Q&A. Moreover, there were 3 parallel sessions including 30 oral presentations covering a wide spectrum of research topics, with presenters from various countries in Asia, Europe, and South America. Each presentation lasted for 15 minutes including 3 minutes Q&A.

I would like to express my sincere gratitude to all in the Technical Committee who have reviewed the papers and developed a very interesting Conference Program, as well as to the keynote speakers. I would also like to thank the organizing committee, the organizing secretariat, as well as all the authors for contributing with their latest research to the success of this conference.

The proceeding of ICBMM 2020 includes a selection of papers submitted to the conference by various authors from different universities, research institutes and industries. All the papers were subjected to peer-review by conference committee members and international reviewers. They were selected for their quality and relevance to the conference.

ICBMM 2020 aimed to provide an excellent avenue for researchers, engineers, and academics to share their research findings and to build networks for further collaborative research in their respective areas. I hope this volume will prove useful for researchers and practitioners in developing and applying new technologies and processes in the field of building materials and materials engineering.

Prof. Carlos Chastre
NOVA University of Lisbon, Portugal

IOP Conf. Series: Materials Science and Engineering 1054 (2021) 011001 doi:10.1088/1757-899X/1054/1/011001

Conference Committee

Conference Chair
Prof. Carlos Chastre, NOVA University of Lisbon, Portugal

Program Chairs
Assoc. Prof. Paulo Mendonça, University of Minho, Portugal
Prof. WONG Kwai Kwan, University of Lyon, France
Prof. Nuno Dinis Cortiços, University of Lisbon, Portugal

Publicity Chairs
Prof. Rudolf Hela, Brno University of Technology, Czech Republic
Prof. Han-Yong Jeon, Inha University, South Korea
Assoc. Prof. Marina Rynkovskaya, Peoples' Friendship University of Russia (RUDN University), Russia

Technical Committee
Prof. S. Joseph Antony, University of Leeds, UK
Prof. Alessandro Pegoretti, University of Trento, Italy
Dr. José Rodríguez Barboza, Universidad Peruana de Ciencias Aplicadas, Perú
Prof. Yao-Ming Hong, Nanhua University, Taiwan
Prof. Seunghee Park ,Sungkyunkwan University , Korea
Prof. Ruby Mejía de Gutiérrez， Universidad del Valle, Colombia
Prof. Zygmunt Meyer,West Pomeranian University of Technology in Szczecin/Department of Geotechnics, Szczecin, Poland
Assoc. Prof. Ali Karrech, University of Western Australia, Australia
Assoc. Prof. Kong Fah Tee, University of Greenwich, UK
Assoc. Prof. Oscar Eugenio Bellini, Politecnico di Milano, Milan, Italy
Assoc. Prof. Adhikari Rajendra Singh, Politecnico di Milano, Milan, Italy
Assoc. Prof. Josè Luìs Barroso de Aguiar, University of Minho, Portugal
Assoc. Prof. Manny Anthony Martin Taguba , National University - Manila , Philippines
Assoc. Prof. Allan R. Alzona, National University - Manila, Philippines
Assoc. Prof. Khairunisa Muthusamy, Universiti Malaysia Pahang, Malaysia
Assoc. Prof. Mohammad Arif Kamal, Aligarh Muslim University, India
Assoc. Prof. Didier Talamona, Nazarbayev University, Kazakhstan
Assoc.Prof. Hosein Naderpour,Semnan University, Iran
Dr. Hui Li, Washington State University, USA
Dr. Donghee Ko,Shimizu Corporation, Tokyo, Japan
Dr.Eduardo Sánchez Caballero,Universidad de Sevilla,Spain
Dr. Ana I. Almerich Chulia, Universitat Politecnica De Valencia, Spain
Dr. Fredy Huamán Mamani, Universidad Católica San Pablo, Perú
Dr. Kun-Chi Wang,Chao Yang University of Technology, Taiwan

IOP Conf. Series: Materials Science and Engineering 1054 (2021) 011001 doi:10.1088/1757-899X/1054/1/011001

IOP Conf. Series: Materials Science and Engineering 1054 (2021) 011002 doi:10.1088/1757-899X/1054/1/011002

Peer review declaration

All papers published in this volume of *IOP Conference Series: Materials Science and Engineering* have been peer reviewed through processes administered by the Editors. Reviews were conducted by expert referees to the professional and scientific standards expected of a proceedings journal published by IOP Publishing.

- **Type of peer review:** Double-blind
- **Conference submission management system:** iConf Conference System
- **Number of submissions received:** 34 (29 full papers + 5 abstracts)
- **Number of submissions sent for review:** 34
- **Number of submissions accepted:** 16
- **Acceptance Rate (Number of Submissions Accepted / Number of Submissions Received X 100):** 55%
- **Average number of reviews per paper:** 2
- **Total number of reviewers involved:** 25
- **Any additional info on review process:**

All received submissions were subjected to double-blind peer reviews, who are expert or have been experiencing in the related field for years. The accepted papers must be revised according to the referees' comments and suggestions, before inclusion in the conference proceedings. Manuscripts submitted should be original & unpublished and written in accordance with the standard IOP paper template.

- **Contact person for queries:** Ms. Elva Zhang (icbmm@cbees.net)

IOP Conf. Series: Materials Science and Engineering 1054 (2021) 012001 doi:10.1088/1757-899X/1054/1/012001

Mechanical Characterization of New Geopolymeric Materials Based on Mining Tailings and Rice Husk Ash

F.A. Huamán-Mamani[1*], D.L. Mayta-Ponce[1] and G.P. Rodríguez-Guillén[1]

[1]Universidad Católica San Pablo, Arequipa, Perú, Campus Campiña Paisajista s/n Quinta Vivanco, Barrio de San Lázaro, Arequipa, Perú

*fhuaman@ucsp.edu.pe

Abstract. This work presents the results of the thermomechanical evaluation of geopolymeric concrete fabricated from mining tailings, rice husk ash and fine sand. Ten types of geopolymeric concrete were studied and the relationship between the initial volumetric concentrations of the components in the mixtures and the maximum resistance in uniaxial compression under conditions of variable temperature (between ambient and 600 ºC) was analyzed. The results revealed that increases in the concentration of mining tailings and fine sand lead to an increase in the value of the maximum mechanical resistance, in contrast, the increase in the concentration of rice husk ash led to a reduction in the value of the maximum mechanical resistance. Furthermore, increases in test temperature, up to 500 °C, led to systematic increases in maximum mechanical strength. Finally, the geopolymeric concretes presented a brittle-ductile transition between 500 and 600 °C showing only a ductile behavior when tested at 600 °C and only brittle up to test temperatures of 500 °C.

1. Introduction

Geopolymers are a new class of inorganic synthetic materials manufactured by geopolymerization of aluminosilicates from: industrial solid waste, calcined clays, natural minerals, among others, and an alkaline activator in aqueous solution [1]. In recent years, geopolymers are strongly attracting the attention of the international scientific community, mainly due to the variety of applications in which they can be used, derived from their excellent fire resistance, low density, low cost, ease of fabrication, stability. Thermo-chemistry and eco-friendly synthesis [1,2]. For some years geopolymers have been considered revolutionary for their possible use as a replacement for traditional Portland cement [3,4]. Several works have revealed that geopolymeric concrete has the ability to develop up to 70% compressive strength within the first 4 hours of curing at an appropriate temperature [5], in contrast to its Portland cement counterpart that takes several weeks. On the other hand, geopolymeric concrete shows very little shrinkage due to drying; it was found that after one year it presents between 5 to 7 times less shrinkage than conventional Portland cement concrete [6]. Kong et al. [7] studied the mechanical response of Portland cement and geopolymer pastes after being subjected to high temperatures, finding that while at 800 °C geopolymer paste improves its resistance to compression, Portland cement paste, at 400 °C, lost all its residual resistance, this loss of resistance of the Portland cement paste was attributed to the decomposition of $Ca(OH)_2$ that occurs at approximately 400 °C.

On the other hand, the generation and accumulation of large amounts of mining tailings as a consequence of mining exploitation constitutes one of the main environmental problems that the Arequipa region (Peru) supports, in addition, currently there are no methodologies for the treatment and reuse of mining tailings, which further compounds their negative environmental impact [8-10].

Therefore, in this work, several mixtures of geopolymeric concretes manufactured from mining tailings, rice husk ash and fine sand were studied, for their potential use in replacement of conventional Portland cement concretes.

2. Materials and methods

2.1. Fabrication of geopolymeric concrete

Geopolymer concrete (GCs) were fabricated by mixing mining tailings (MT), rice husk ash (RHA), fine sand (FS) and Na(OH) solution (9 molar). MT was provided by the metallurgical company CEPROMET SAC and the RHA was obtained by calcining rice husk at 300 ºC for 3 hours. Ten types of GCs were fabricated, whose volumetric concentrations are shown in table 1.

Table 1. Matrix of mixtures for the fabrication of GCs.

Sample	Volume (%)		
	MT	RHA	FS
GC-MT20.0RHA10.0	20.0	10.0	70.0
GC-MT20.0RHA6.7	20.0	6.7	73.3
GC-MT20.0RHA3.3	20.0	3.3	76.7
GC-MT20.0RHA0.0	20.0	0.0	80.0
GC-MT23.3RHA6.7	23.3	6.7	70.0
GC-MT23.3RHA3.3	23.3	3.3	73.3
GC-MT23.3RHA0.0	23.3	0.0	76.7
GC-MT26.7RHA3.3	26.7	3.3	70.0
GC-MT26.7RHA0.0	26.7	0.0	73.3
GC-MT30.0RHA0.0	30.0	0.0	70.0

MT, RHA and FS were ground separately and then sieved by ASTM No. 140 mesh (106 μm). For each type of GC, and based on 12g of mixture, adequate amounts of raw material were measured (according to table 1), the dry powders were mixed for 5 minutes, then 5 ml of Na(OH) solution was added, mixing continued until a homogeneous and workable paste was obtained. All geopolymeric pastes were compacted in a 20 mm diameter steel cylindrical mold for 5 minutes and at 30 MPa. The pressed cylindrical samples were placed in airtight bags and then heat-cured at 60 °C for 48 hours.

2.2. Physical, structural, microstructural and mechanical characterization

The real density of the raw material was determined by helium pycnometry (Functional Characterization Laboratory, CITIUS, University of Seville, Spain). X-ray diffraction analysis was performed on the starting powders, using a Bruker D8 Advance A25 X-ray diffractometer with Cu Kα radiation and Ni filter (X-Ray Laboratory, CITIUS, University of Seville, Spain). Microstructural characterization was performed by scanning electron microscopy (SEM) (Microscopy Laboratory, CITIUS, University of Seville, Spain). The thermomechanical characterization of the CGs was performed on 5x5x10 mm parallelepipeds and using a MICROTEST EM1/50/FR universal testing machine (Spain). The tests were carried out at temperatures between ambient and 600 ºC, in an air atmosphere and with a constant compression rate of 0.05 mm/min.

3. Results and discussion

3.1 Density, phases and morphology

The real average density found for FS, MT and RHA was 3.07, 2.83 and 2.23 g/cm^3, respectively. Regarding the structural analysis, diffraction peaks of up to five phases were identified in FS: Albite ($Na(Si_3Al)O_8$), Gypsum ($CaSO_4*2H_2O$), Quartz (SiO_2), Potassium Magnesium Aluminum Fluoride Silicate ($KMg(Si_3Al)O_{10}F_2$) and Fluoro-edenite ($NaCa_2Mg_5Si_7AlO_{22}F_2$), nine-phase diffraction peaks in MT: Muscovite (KAl_2Si_3), Quartz (SiO_2), Orthoclase ($KAlSi_3O_8$), Pyrite (FeS_2), Calcite ($CaCO_3$),

Sulfur (S), Actinolite ($Ca_2(Mg,Fe_{+2})_5Si_8O_{22}(OH)_2$), Hydrogen Arsenate Hydrate (As_2O_5*$4H_2O$) and Arsenolite (As_2O_3); and a completely amorphous baseline for RHA, without the presence of diffraction peaks.

Figure 1 shows SEM micrographs of powder from FS (figure 1(a)), MT (figure 1(b)) and RHA (figure 1(c)), significant differences were found in particle sizes, those of FS being those of larger, it is also observed that the shape of the FS particles are mostly angular and those of MT and RHA rounded.

Figure 1. SEM micrographs of raw material powder (a) FS, (b) MT and (c) RHA.

Figures 2(a) and 2(b) show SEM micrographs of polished surfaces of GCs with 70 and 80 vol.% FS, respectively. For all the GCs studied, two differentiated phases were found: a continuous phase that corresponds to the geopolymer (mixture of MT and RHA activated alkaline) and another dispersed phase of FS particles located within the continuous phase of geopolymer. Greater grain cohesion has been observed in GCs with 70 vol.% FS (figure 2(a)), compared to poor grain cohesion of the GC with 80 vol.% FS (figure 2(b)).

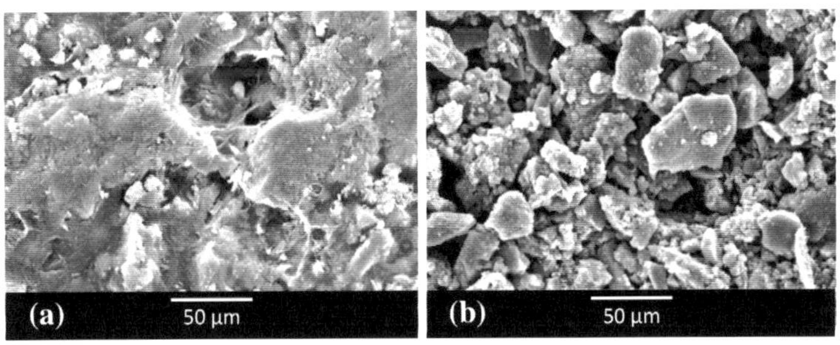

Figure 2. SEM micrographs of polished surfaces of GCs with (a) 70 vol.% and (b) 80 vol.% FS.

3.2. Compressive strength

Figure 3 shows curves of maximum compression stress vs. test temperature. A systematic increase in mechanical resistance could be seen by increasing the temperature up to 500 °C, from this temperature an evolution in mechanical behavior occurs, which suggests the activation of microscopic mechanisms of plastic deformation, which led the samples to enter in a state of steady creep with mechanical strengths below 8 MPa in all cases.

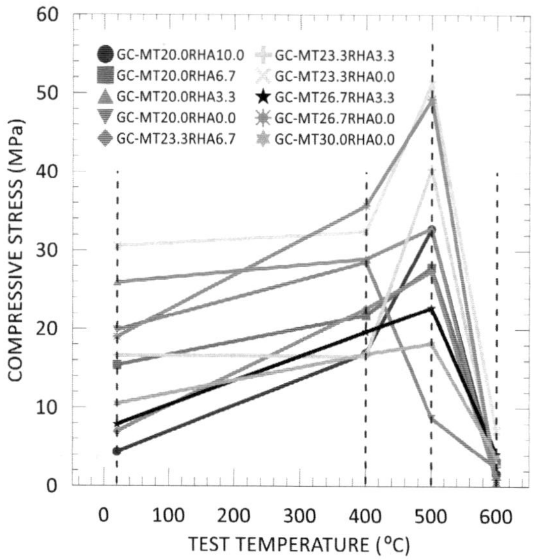

Figure 3. Compressive stress as a function of test temperature.

Figure 4 shows the relationship between the maximum mechanical response in compression and the volumetric fraction of FS, when the volumetric fraction of MT is kept constant. An increase in mechanical resistance is observed as the volumetric fraction of FS increases, this relationship is valid up to approximately 76.7 vol% of FS, since from this volumetric fraction the mechanical resistance begins to decrease, this decrease in mechanical resistance is It can be attributed to the limited cohesion of the sand particles due to the excessive volumetric concentration of fine sand particles in relation to the volumetric concentration of binder phase (MT and RHA).

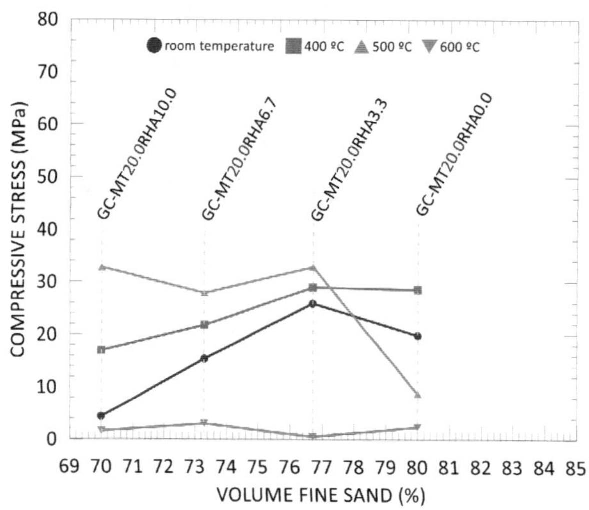

Figure 4. Compressive stress as a function of volume fine sand and keeping the volume of mining tailings constant.

4. Conclusions

Geopolymeric concretes were successfully fabricated by geopolymerization of mining tailings powder, rice husk ash, fine sand, and an aqueous solution of sodium hydroxide (9 molar).

Geopolymeric concrete presented two phases: a continuous phase of geopolymer and another dispersed phase of fine sand particles located within the continuous phase of geopolymer.

It was determined that increases in the concentration of mining tailings or fine sand in geopolymeric concretes, leads to increased values of maximum mechanical resistance, in contrast, increases in the concentration of rice husk in geopolymeric concretes, leads to decrease of the values of maximum mechanical resistance.

The increase in temperature in mechanical compression tests, up to 500 ºC, lead to systematic increases in maximum mechanical resistance.

Geopolymeric concrete presented a brittle-ductile transition between 500 and 600 °C, showing ductile behavior when tested at 600 °C and brittle to temperatures below 500 °C.

5. References

[1] Hamdy K. Shehab, Ahmed S. Eisa, Ahmed M. Wahba, Mechanical properties of fly ash based geopolymer concrete with full and partial cement replacement, *Construction and Building Materials*, Volume 126, 2016, pp 560-565.

[2] Bo Wei, Yimin Zhang, Shenxu Bao, Preparation of geopolymers from vanadium tailings by mechanical activation, *Construction and Building Materials*, Volume 145, 2017, pp 236-242.

[3] F.N. Okoye, Binder: A veritable alternative to Portland cement, *Materials Today: Proceedings*, Volume 4, Issue 4, Part E, 2017, pp 5599-5604.

[4] Robert J. Flatt, Nicolas Roussel, Christopher R. Cheeseman, Concrete: An ecomaterial that needs to be improved, *Journal of the European Ceramic Society*, Volume 32, Issue 11, 2012, pp 2787-2798.

[5] Davidovits J., Properties of Geopolymer Cements. *First International Conference on Alkaline Cements and Concretes* 1994, pp 131-149.

[6] Wallah SE, Drying Shrinkage of Heated-Cured Fly Ash-Based Geopolymer Concrete, *Modern Applied Science* 2009;3(12), pp 14-21.

[7] Kong DLY and Sanjayan JG., Effect of elevated temperature son geopolymer paste, mortar and concrete. *Cement and Concrete Research* 2010;40(2), pp 334–339.

[8] Mónica Moreno-Brush, Johan Rydberg, Nadia Gamboa, Ilse Storch, Harald Biester, Is mercury from small-scale gold mining prevalent in the southeastern Peruvian Amazon?, *Environmental Pollution*, Volume 218, 2016, pp 150-159.

[9] Joni Safaat Adiansyah, Michele Rosano, Sue Vink, Greg Keir, Jason R. Stokes, Synergising water and energy requirements to improve sustainability performance in mine tailings management, *Journal of Cleaner Production*, Volume 133, 2016, pp 5-17.

[10] Joni Safaat Adiansyah, Michele Rosano, Sue Vink, Greg Keir, A framework for a sustainable approach to mine tailings management: disposal strategies, *Journal of Cleaner Production*, Volume 108, Part A, 2015, pp 1050-1062.

Acknowledgements

This work was supported by the project no. 143-2015-FONDECYT-DE, Consejo Nacional de Ciencia, Tecnología e Innovación Tecnológica – CONCYTEC, Perú.

IOP Conf. Series: Materials Science and Engineering 1054 (2021) 012002 doi:10.1088/1757-899X/1054/1/012002

Fabrication and Evaluation of the Mechanical Behavior of Geopolymer Compounds Using Waste from the Mining and Construction Industry

FA Huamán-Mamani[1*], DL Mayta-Ponce[1] and GP Rodríguez-Guillén[1]

[1]Universidad Católica San Pablo, Arequipa, Perú, Campus Campiña Paisajista s/n Quinta Vivanco, Barrio de San Lázaro, Arequipa, Perú

*fhuaman@ucsp.edu.pe

Abstract. Geopolymers are lately being considered a group of revolutionary materials, due to their good mechanical properties, chemical stability, fire resistance and diversity of applications where they can be used. For the synthesis of geopolymers, a great variety of types of natural and artificial raw materials (residues from other industries) can be used, which allows obtaining materials with very specific applications. Two of the most important industries in Peru are also those that generate greater environmental degradation, mainly due to the waste they generate. The mining industry has a negative impact on the environment, generating a large amount of inorganic waste, while the construction industry does the same, generating large amounts of demolition waste. Both, mining and demolition waste, constitute a serious environmental problem, since currently they are only deposited without any alternative use. Therefore, our research proposes the use of geopolymeric technology for the use of mining and demolition waste in the manufacture of geopolymeric concrete with mechanical strengths similar to that of Portland cement concrete.

1. Introducción

The action of alkaline compounds on an aluminosilicate source can generate three-dimensional cross-linked structures of SiO_4 and AlO_4, which constitute a new type of inorganic materials called geopolymers [1,2]. Geopolymers are materials that in recent years have been attracting the attention of the scientific community, because they have good thermal and chemical stability, good mechanical resistance and are easy to manufacture [1,3]. The construction industry has a particular interest in the development of these materials, because geopolymers have been shown to be excellent candidates to replace traditional Portland cement [4,5].

Concrete is a material widely used in contemporary construction, it is relatively cheap and durable, but it has an associated negative impact on the environment [6]. For the manufacture of geopolymers, natural raw materials (clays, for example) and industrialized (waste from other industries) are used. Several studies have shown that geopolymeric concrete has the ability to develop up to 70% compressive strength within the first four hours of cure [7], compared to its Portland cement counterpart that takes several weeks.

We should point out that there is currently great concern about the accumulation of mining tailings as a consequence of mining in Peru. Additionally, currently there are no adequate methodologies for the treatment or reuse of mining tailings, which further aggravates their environmental impact [8,9]. On the other hand, infrastructure repairs generate a significant amount of Portland cement concrete waste, in addition to the above, the difficulty of finding a suitable place for the final disposal of this waste [10,11]. Therefore, this work aimed to obtain geopolymeric concrete with good mechanical properties, similar to that of its Portland cement counterparts, from a mixture of mining tailings and recycled Portland cement mortar, without the addition of Portland cement.

2. Materials and methods

2.1. *Fabrication of geopolymeric concrete*

Geopolymer concrete (GCs) were obtained by mixing controlled amounts of mining inorganic waste (MIW) powder, Portland cement recycled mortar (PCRM) powder and a Na(OH) solution (9 molar). Five types of mixtures were prepared with concentrations shown in table 1.

Table 1. Compositional matrix of mixtures for studied GCs.

Sample	Vol. %		Wt. %		Wt. (g)	
	MIW	PCRM	MIW	PCRM	MIW	PCRM
MIW(30)	30	70	32	68	3.80	8.20
MIW(40)	40	60	42	58	5.02	6.98
MIW(50)	50	50	52	48	6.23	5.77
MIW(60)	60	40	62	38	7.42	4.58
MIW(70)	70	30	72	28	8.59	3.41

For the determination of mass concentrations of GCs, real density values for MIW and PCRM of 2.7 and 2.5 g/cm^3, respectively, were considered. The fabrication of GCs began with the collection of the raw material. MIW was provided by the company CEPROMET SAC (dedicated to gold metallurgy) and the PCRM was obtained from a demolition waste collection center located in Arequipa, Peru.

MIW and PCRM were crushed and sieved by ASTM No. 140 sieve, obtaining a fine powder with particle sizes of less than 106 μm. Based on table 1, MIW and PCRM masses were measured, considering 12 g as the total mixture mass for each sample, in parallel, a Na(OH) solution (9 molar) was prepared. MIW and PCRM powders were dry mixed for 5 minutes and then 3 ml of Na(OH) solution was added, mixed for an additional 5 minutes to form a homogeneous and workable paste. All pastes were uniaxially pressed in a 20 mm diameter steel mold at 20 MPa and for 5 minutes. The samples obtained were placed in hermetic bags for curing at room temperature for 24 hours and then brought to 80ºC for 24 hours.

Each cylindrical sample of GCs fabricated was cut into approximately 5x5x10 mm parallelepiped samples. For cutting the parallelepipeds, a low-speed precision cutter fitted with a diamond-edge cutting disc was used. The parallelepipeds obtained were used for the physical, morphological and mechanical characterization tests.

2.2. *Physical, morphological and mechanical characterization of GCs*

The physical characterization consisted of determining its geometric density from the measurement of the parallelepiped dimensions and masses of each type of GC studied. Morphological characterization began with the metallographic preparation of a face of a parallelepiped of a certain type of GC, this preparation consisted of two stages: (i) roughing with SiC abrasive paper (# 600, # 800 and # 1200) and (ii) polishing with 6, 3 and 1 μm particle size diamond paste. After polishing, the surfaces were cleaned with distilled water and placed in the sun to evaporate the remaining moisture. Finally, the polished, clean and dry surfaces were observed in an optical microscope in reflection mode. The mechanical characterization consisted of uniaxial compression tests at room temperature and at a constant compression speed of 0.05 mm/min. Six parallelepipeds were tested for each type of GC. All mechanical tests were performed after 28 days of curing.

3. Results and discussion

3.1. Physical and morphological characterization

The average geometric density for all GCs was 1.8 g/cm^3, no relationship was found between the initial volumetric composition of the different components (MIW and PCRM) and the geometric density of GCs. The morphology found in all the GCs consisted of a continuous geopolymer phase (continuous light phase in figure 1) with dark particles dispersed in the continuous matrix, which corresponds to fine sand (FS) particles from the PCRM. It should be noted that the PCRM typically consists of a mixture of Portland cement (PC) and FS, therefore, by crushing and sieving the PCRM, mainly PC with a minimal amount of FS particles is obtained. This is due to the fact that FS has a particle size greater than 106 μm and PC particles have smaller sizes. On the other hand, it was verified that the MIW had a particle size of less than 106 μm.

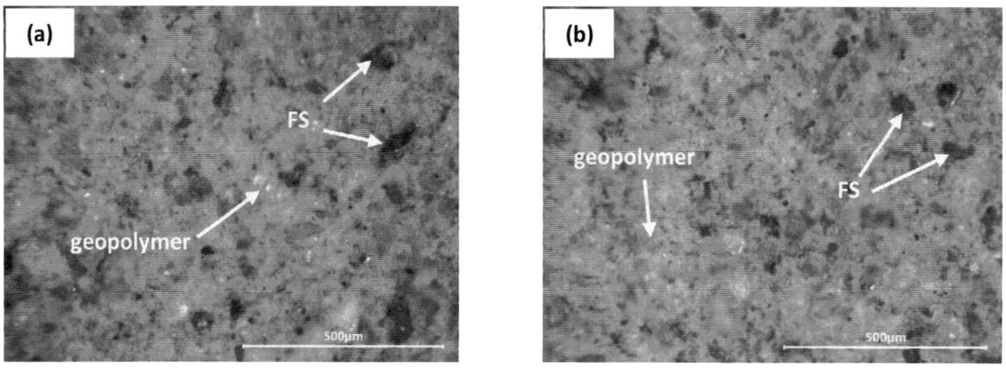

Figure 1. Micrographs by optical microscopy of (a) MIW(30), (b) MIW(70).

3.2. Mechanical characterization

Figure 2 presents stress vs. strain curves for all the GCs, in all cases a fragile behavior was found, with higher maximum resistance values in the GCs with the highest concentration of PCRM (MIW (30) and MIW (40)) (figure 2(f)). From concentrations higher than 50% of PCRM in GCs, the maximum resistance values were kept below 20 MPa. GCs with 60 vol.% of MIW presented a significant reduction in maximum resistance, compared to GCs with 50 vol.% of MIW, this reduction was accompanied by an increase in deformation, going from 2% in MIW(50) to 4% in MIW(60). Additionally, the mechanical response of the GC MIW(70), in maximum resistance and deformation, was very similar to that of the GC MIW (50). Figure 2(f) presents a comparison between the maximum resistance values of the five types of GCs studied, for this comparison more representative stress vs. strain curves were chosen.

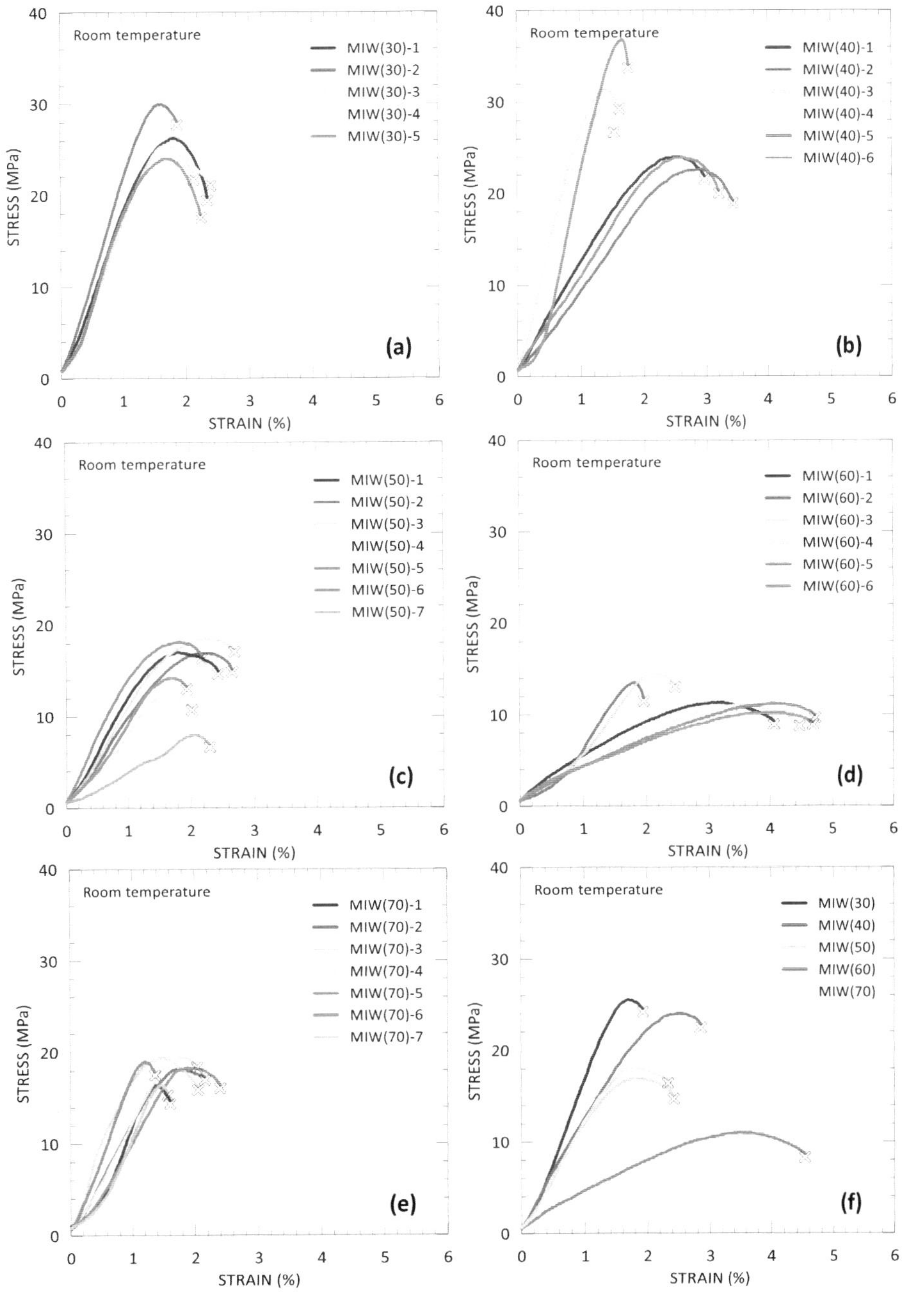

Figure 2. Stress vs. strain curves for all geopolymeric concretes studied.

4. Conclusions

Geopolymeric concrete has been successfully fabricated from the geopolymerization of mining inorganic waste powder, Portland cement recycled mortar and a Na(OH) solution 9 molar. No relationship was found between the geometric density of the geopolymeric concretes and the volumetric fractions of the precursor components. In general, the average geometric density of all the studied geopolymeric concretes was 1.8 g/cm^3.

The morphology found in geopolymeric concretes consisted of two different phases, one continuous of light contrast (geopolymer) and the other of dispersed particles (fine sand) of dark contrast, located within the continuous phase of geopolymer.

A direct correlation was identified between the volumetric fraction of Portland cement recycled mortar in the GCs and their corresponding maximum mechanical strength.

The geopolymeric concretes studied presented the microscopic fracture mechanisms typical of ceramic materials, which include nucleation, propagation and connection of cracks until the final fracture.

5. References

[1] Bo Wei, Yimin Zhang, Shenxu Bao. Preparation of geopolymers from vanadium tailings by mechanical activation. *Construction and Building Materials*. vol. 145, 2017. pp 236-242.

[2] Márlon A. Longhi, Erich D. Rodríguez, Susan A. Bernal. Valorisation of a kaolin mining waste for the production of geopolymers. *Journal of Cleaner Production*. vol. 115, 2016, pp 265-272.

[3] Jingkun Yuan, Peigang He, Peifeng Zhang, Dechang Jia, Delong Cai, Zhihua Yang, Xiaoming Duan, Shengjin Wang, Yu Zhou. Novel geopolymer based composites reinforced with stainless steel mesh and chromium powder. *Construction and Building Materials*. vol. 150, 2017, pp 89-94.

[4] J. Davidovits, Geopolymer-inorganic polymeric new materials. *Journal of Thermal Analysis*, 37 (8) (1991) pp 1633–1656.

[5] B. Singh, G. Ishwarya, M. Gupta, S.K. Bhattacharyya. Geopolymer concrete: A review of some recent developments. *Construction and Building Materials*. vol. 85, 2015, pp 78-90.

[6] Hrvoje Mikulčić, Jiří Jaromír Klemeš, Milan Vujanović, Krzysztof Urbaniec, Neven Duić. Reducing greenhouse gasses emissions by fostering the deployment of alternative raw materials and energy sources in the cleaner cement manufacturing process. *Journal of Cleaner Production*. vol. 136, 2016, pp 119-132.

[7] Nicoletta Toniolo, Aldo R. Boccaccini. Fly ash-based geopolymers containing added silicate waste. A review. *Ceramics International*. vol. 43, 2017, pp 14545-14551.

[8] Mónica Moreno-Brush, Johan Rydberg, Nadia Gamboa, Ilse Storch, Harald Biester. Is mercury from small-scale gold mining prevalent in the southeastern Peruvian Amazon?. *Environmental Pollution*. vol. 218, 2016, pp 150-159.

[9] Joni Safaat Adiansyah, Michele Rosano, Sue Vink, Greg Keir, Jason R. Stokes. Synergising water and energy requirements to improve sustainability performance in mine tailings management. *Journal of Cleaner Production*. vol. 133, 2016, pp 5-17.

[10] Hack DR, Bryan DP. Aggregates. Industrial Minerals and Rocks, Kogel EK, Trivedi NC, Barker JM, Krukowski ST (eds), 7th Edition, Littleton, Colorado, *Society for Mining, Metallurgy and Exploration*; 2006. p. 1105-1119.

[11] Arulrajah A, Piratheepan J, Disfani MM, Bo MW. Geotechnical and geoenvironmental properties of recycled construction and demolition materials in pavement subbase applications. *J Mater Civ Eng* 2013; 25(8), pp 1077–1088.

Acknowledgements

This work was supported by the project no. 143-2015-FONDECYT-DE, Consejo Nacional de Ciencia, Tecnología e Innovación Tecnológica – CONCYTEC, Perú.

Influence of the Activating Solution and Aggregates in the Physical and Mechanical Properties of Volcanic Ash Based Geopolymer Mortars

Nataly Calderón[1], María Vargas[2], Jonathan Almirón[3], Asunción Bautista[4], Francisco Velasco[5] and Danny Tupayachy-Quispe[6*]

1,3 Msc. Ciencia y Tecnología de los Materiales, Escuela Profesional de Ingeniería Ambiental, Facultad de Ingeniería de Procesos, Universidad Nacional de San Agustín de Arequipa. Av. Independencia 400, Cercado – Arequipa (Perú).
4,5 Materials Science and Engineering Department, IAAB, Universidad Carlos III de Madrid. Avda. Universidad N° 30. 28911 - Leganés, Madrid (Spain).
2,6 Universidad Católica de Santa María, Laboratorio de Ciencia de los Materiales, Samuel Velarde 320, Umacollo, Arequipa, Perú

*dtupayachy@ucsm.edu.pe

Abstract. The physical, mechanical and microstructural properties of volcanic ash based geopolymer mortars by alkaline activation (with NaOH and Na(OH) + Na_2SiO_3 solutions) using different aggregates (Peruvian sand and standard sand) were studied. For the synthesis of geopolymers, a liquid:solid ratio of 0.1 and an ash:sand ratio of 1:1 were selected, and 25% of sodium silicate was added to the Na(OH) solution. The samples were compacted, toughened and cured at 80°C for 48 h. The dissolution of volcanic ash particles and the formation of the geopolymeric matrix increase with Na(OH)+Na_2SiO_3 solution and standard sand while the local sand promotes better geopolymerization with NaOH solution. The compression strength of volcanic ash based geopolymer mortar reported values from 18.93 MPa to 23.28 MPa at 6 months of curing, which determines the use of these geopolymers as a construction material.

1. Introduction

Geopolymers, known as inorganic polymers or zeolite precursors, are aluminosilicate materials with attractive properties causing a variety of studies to replace Portland cement because the production of Portland cement results in an excessive carbon dioxide emissions (which are 6-7% of the worldwide emissions) [1]. These inorganic polymers have good physical and mechanical properties, such as stability at high temperatures and high compression strength [2]. These properties rely on a variety of parameters, such as the type of precursor materials, mineralogical and chemical compositions of the aluminosilicate sources, temperature and curing time, alkaline composition of the solution, the liquid:solid and aggregate ratios [3]. Additionally, a variety of studies has revealed that volcanic ashes, metakaolin and different aluminosilicates can be used as precursor materials for the synthesis of geopolymers.

Volcanic ashes contain a large amount of silica and alumina which make them suitable for the production of geopolymers [4]. These vitreous materials also have a certain amount of innocuous and inactive minerals such as magnetite, quartz and mica, leading into an important challenge to understand the reaction of phases during and after the hydration process. Nowadays, there have been developed studies concerning volcanic ashes as a replacement of Portland cement. Volcanic ashes are widely available in the south area of Peru. The Ubinas volcano is considered as the most active volcano in Peru because of its 25 events of high fumarolic activities followed by ashes and rocks emissions since XVI century [5]. The distribution of ashes around the volcano, determined by the influence of winds, allowed the formation of stockpiles, which are reachable and it can contribute into and easy extraction followed by economic and environmental benefits to the country.

The current investigation has studied the physical, mechanical, mineralogical and microstructural properties of volcanic ashes based geopolymer mortars, using different types of aggregates such as local sand (LS) and standard sand (SS). The geopolymer mortars were under evaluation for 6 months, with an ash:sand ratio of 1:1. The alkaline activation had a liquid:solid ratio of 0.1, with sodium hydroxide (NaOH) and sodium silicate + sodium hydroxide ($NaOH+Na_2SiO_3$) solutions, this last in a concentration of 75%-25% (w/w)(), and both solutions were in 12 M concentration.

2. Experimental procedure

2.1. Materials

Volcanic ashes (VA) were collected from Ubinas volcano, whose geographical coordinates are 297608 of latitudes, 8196158 length and 4600 meter above sea level (Moquegua, Peru). The aggregate of local sand (LS) was from Arequipa, and the aggregate of standard sand (SS) was standardized CEN-NORMSAND (according to DIN EN 196-1 standard), with a particle size ranging between 0.08 and 2 mm. The volcanic ashes were ground in a planetary ball mill Fritsch 6S for 25 min at 450 rpm in order to get fine particles (<75 μm). The volcanic ashes and aggregates were dried in Memmert Universital Oven UN55 at 105 °C for 6 h. Portland cement from Yura IP (YC) was used for comparison to the geopolymer mortars.

2.2. Method of synthesis of geopolymeric mortars

The synthesis of geopolymer mortars was made by alkali activation. Volcanic ashes and aggregates were in a ratio of 1:1. They were dissolved in an alkaline solution with a liquid:solid ratio of 0.1. The solutions of (NaOH) and ($NaOH+Na_2SiO_3$) were in 12 M concentration, and $NaOH+Na_2SiO_3$ was in a proportion of 75%-25%. The samples were mixed according to the UNE-196-3 standard, dumped into cylindrical molds of 28 mm diameter and followed by compression at 15 MPa. The samples were covered with a plastic film to be cured at 80°C for 48 h in a Memmert Universal Oven UN-50. After curing, the samples were placed in a Memmert Incubator IF-110 at room temperature and high relative humidity (>90%) until the compression test day (7, 14, 28, 90 and 180 days).

3. Characterization techniques of materials and geopolymeric mortars

3.1. X-ray Fluorescence Spectroscopy (XRF)

The chemical composition of precursor materials was determined by an X-ray Fluorescence Spectrometer (XRF) Spectro Xepos.

3.2. X-ray Diffraction (XRD)

The mineralogical composition of precursor materials and geopolymer mortars was analyzed by X-ray diffraction, Bruker (DRX) D8, with a radiation or CuKα (λ = 0, 1542 nm), voltage of 40 kV and current of 40 mA, the angle 2θ was in arrange of 10° to 80° with a scan speed of 2 °/min.

3.3. Absorption, and bulk density

The bulk density was determined by gravimetric method (UNE-EN 1015-11:1999), Archimedes method (ASTM C-20) and the absorption analysis by ASTM C-642.

3.4. Compression strength
The compression strength of geopolymer mortars was determined at 7, 14, 28, 90 and 180 days of curing according to UNE-EN 12390-3:2003. It was carried out in a hydraulic compression machine ELE International. Five tests were made for each condition.

3.5. Scanning Electron Microscopy (SEM)
The morphological characteristics of geopolymer mortars were analyzed by a Scanning Electron Microscope Hitachi SU8230 with back scattered electron (BSE). The images were taken at 5kX.

4. Results and discussion

4.1. Materials characterization

4.1.1. X-ray Fluorescence Spectroscopy (XRF). The chemical composition of precursor materials used in this work are shown in table 1. They are mainly composed of SiO_2, Al_2O_3 and Fe_2O_3. The composition is favorable as the values are ranged into the main requirements in precursor materials for the production of geopolymers [6].

Table 1. Chemical composition of volcanic ash, cement and aggregates (wt. %)

	YC	VA	LS	SS
SiO_2	31.68	45.26	45.33	93.90
Al_2O_3	6.32	13.40	14.47	4.04
Fe_2O_3	3.28	6.72	8.03	1.38
CaO	29.86	5.18	5.23	0.58
MgO	0.43	3.58	2.73	0.01
K_2O	1.48	1.48	1.17	-
TiO_2	0.17	1.08	0.74	0.06
P_2O_5	0.17	0.37	0.21	0.02
Na_2O	0.01	0.26	1.44	-
MnO	0.04	0.10	0.08	0.01
LOI[a]	26.56	22.57	20.57	-

[a] Loss on ignition

The hydraulic cement (YC) is formed by 29.86% of calcium oxide (CaO), 31.68% of silicon dioxide (SiO_2), and low quantities of aluminum oxide (Al_2O_3) and iron oxide (Fe_2O_3).

Regarding the chemical composition of volcanic ash (VA), it shows SiO_2/Al_2O_3 molar ratio of 5.73. These components are the major constituent in precursor materials for the production of geopolymers and their values match those reported by Davidovits [7] and Palomo [8]. It is well known that the small amounts of calcium oxide influence significantly in the properties of geopolymer cements. In this case, CaO is 5.18%. The chemical composition confirms that volcanic ashes are predominantly constituted of silica followed by aluminum, iron, calcium and sodium oxides. The amount of (SiO_2+Al_2O_3) %wt is beneficial for the synthesis of geopolymers. Therefore, volcanic ash based geopolymer can be effective [7] and its alumina composition is beneficial for the mitigation of efflorescence [9].

The chemical composition of the aggregates was also analyzed, LS showing 45.33% of SiO_2. On the other hand, SS has a high amount of SiO_2 (93.91%), which is similar to the dune sand reported by Chuah [10] on his studies of the properties of fly ash based geopolymer mortars.

4.1.2. X-ray Diffraction (XRD). The X-ray diffractograms of precursor materials in an angle range of 10° to 80° of 2θ are shown in figure 1. The cement shows the major peaks of tricalcium silicate ($3CaO.SiO_2$) or alite (C_3S), bicalcium silicate ($3CaO.SiO_2$) or belite (C_2S), tricalcium aluminate ($3CaO.Al_2O_3$) or (C_3A), and tetracalcium ferritoaluminate ($4CaO.Al_2O_3.Fe_2O_3$) or (C_4AF). According to other studies, the crystalline system of C_3S and C_2S phases is monoclinic, C_3A is cubic and C_4AF is orthorhombic [11], [12]. There are also another phases such as calcium oxide or lime, magnesium oxide and alkaline sulfates [11].

The XRD of volcanic ashes shows peaks of augite, muscovite, anortite, forsterite, quartz, feldspars, magnetite, sodium diopside, and aluminum diopside [9], [13],[14]. Unlike other studies, this volcanic ash has a prominent peak between angles 50° and 55°, which corresponds to the presence of diopsides. The characterization and classification of volcanic ashes is based on the mineralogical composition. The aggregates show peaks of quartz and alpha quartz, as well as a high amount of calcite [15].

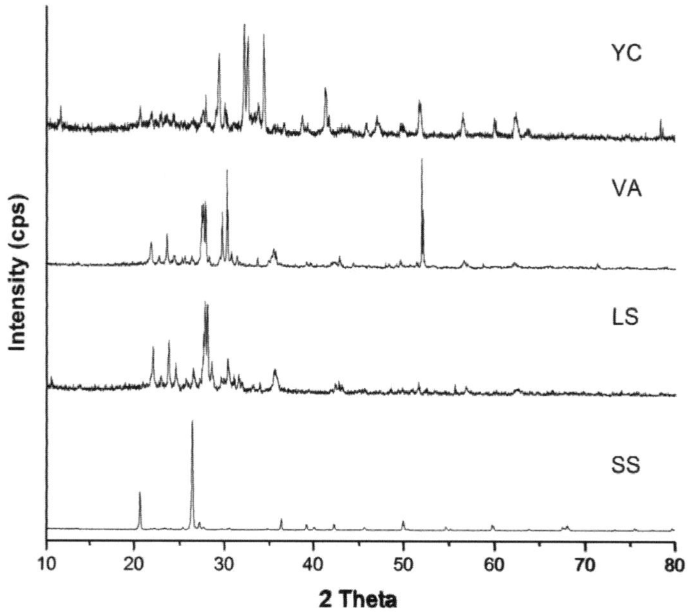

Figure 1. DRX patrons of cement (YC), volcanic ash (VA), local sand (LS), and standard sand (SS).

4.2. Physical and mechanical tests of mortars

4.2.1. Water absorption and bulk density. Table 2 shows the water absorption, Archimedes density and bulk density (%) for samples cured at 28 days. The water absorption varies from 8.54% to 12.96% where samples VA-SS-1 and VA-SS-2 have the lowest values. Although those values are slightly above the reported in other studies like Djobo [9], it is well known that water absorption correlates with the values of apparent porosity which is the result of water that could not react during the geopolymerization. Therefore, the higher water absorption, the higher porosity in mortars. The bulk and Archimedes' densities range from 1.92 to 2.11 g/cm^3 and 1.74 to 2.05 g/cm^3 respectively, showing VA-SS-1 and VA-SS-2 samples the highest values. Density also correlates to the apparent porosity. Porosity decreases when the density increases and vice versa affecting the mechanical properties of mortars [16]. Thus, good results on compression strength are expected for volcanic ash based geopolymer mortars activated with Na(OH)+Na_2SiO_3, better than YC-LS samples.

Additionally, the influence of the aggregates and alkaline solution in samples with LS solution causes higher values of density using NaOH solution, resulting in less porosity. In contrast, samples with SS have higher density values when using NaOH+ Na_2SiO_3 solution, indicating this solution has reacted much better with SS-containing mortars than with LS ones. According to the chemical composition of aggregates, the higher amount of silica contributes to a better polycondensation [13].

Table 2. Absorption, density, and bulk density of volcanic ash based geopolymer mortars

Mortar	Activator	Absorption %	Bulk density (g/cm^3)	Archimedes density (g/cm^3)
YC-LS	H_2O	11.66	1.96	1.84
VA-LS-1	Na(OH)	12.64	1.95	1.80
VA-LS-2	Na(OH)+Na_2SiO_3	12.96	1.92	1.74
VA-SS-1	Na(OH)	10.90	2.02	1.96
VA-SS-2	Na(OH)+Na_2SiO_3	8.54	2.11	2.05

4.2.2. Compression strength. The evolution of the compression strength of volcanic ash based geopolymer mortars with time (from 7 days to 180 days) is reported in table 3. The maximum strength for sample VA-SS-2 is 38,84 MPa at day 7th which decreases over time until 23.28 MPa on day 180. However, this sample with Na(OH) solution has the minimum strength with values of 16.94 MPa, 23.62 MPa and 11.81 MPa at days 7, 90 and 180 respectively. Apart from that, sample VA-LS-1 shows better strengths with Na(OH)+Na_2SiO_3 for each curing time. According to different studies, the gradual decrease on the strength of geopolymer mortars in comparison with cement Portland is accredited to the presence of calcium, porosity or an unfinished geopolymerization [9], [17], [18]. Additionally, the compressive strength can be related to the characteristics of the aggregates and their bind with the polymer gel [19], along with the absorption and density properties [9]. Therefore, it can be inferred that the values of compressive strengths for samples VA-LS are accredited to an unfinished geopolymerization, contrary to the samples VA-SS which may have a complete geopolymerization.

Table 3. Compression strength (MPa) of volcanic ash based geopolymer mortars.

Mortar	Activator	7 days	14 days	28 days	90 days	180 days
VA-LS-1	Na(OH)	27.49±8.91	27.57±6.28	24.57±1.72	23.15±5.12	18.93±1.49
VA-LS-2	Na(OH)+ Na_2SiO_3	17.46±6.69	12.67±2.62	14.15±2.33	16.75±1.08	11.81±1.02
VA-SS-1	Na(OH)	16.94±1.82	14.99±1.88	20.21±7.31	23.62±5.91	11.81±2.81
VA-SS-2	Na(OH)+ Na_2SiO_3	38.84±2.57	31.835.67	26.30±5.67	21.58±8.26	23.28±5.14
YC-LS	H_2O	27.45±3.44	28.03±4.77	29.36±2.57	27.36±0.56	25.10±7.96

4.3. Mineralogical characterization of mortars

4.3.1. X-ray Diffraction (XRD). X-ray diffraction patterns of volcanic ash based mortars are shown in Figure 2. The diffraction patterns for samples YC-LS and VA-LS are similar and there are no variations. On the other hand the diffraction pattern for sample VA-SS shows a slightly shift of its major peak located at the angle of 28°. As mentioned in the studies developed by Baenla [20], minerals present in volcanic ashes have not been affected by the alkaline activation except muscovite, arcanite and calcite, suggesting an incomplete dissolution of crystalline phases in precursor materials [14]. Also, it is worth mentioning that the dissolution of muscovite is marked by the decrease of its peak followed by an increase of bents between angles 18° and 30° in the diffraction patterns of mortars compared to the volcanic ash. However, the diffraction pattern of volcanic ash based geopolymer mortars also shows the appearance of sodium aluminum silicate hydroxide hydrate $Na_8(AlSiO_4)_6(OH)_2 4H_2O$ at angles between 20° and 25°. This can be related to the nature of the precursor materials, curing temperature and the concentration/composition of the alkaline solution [14].

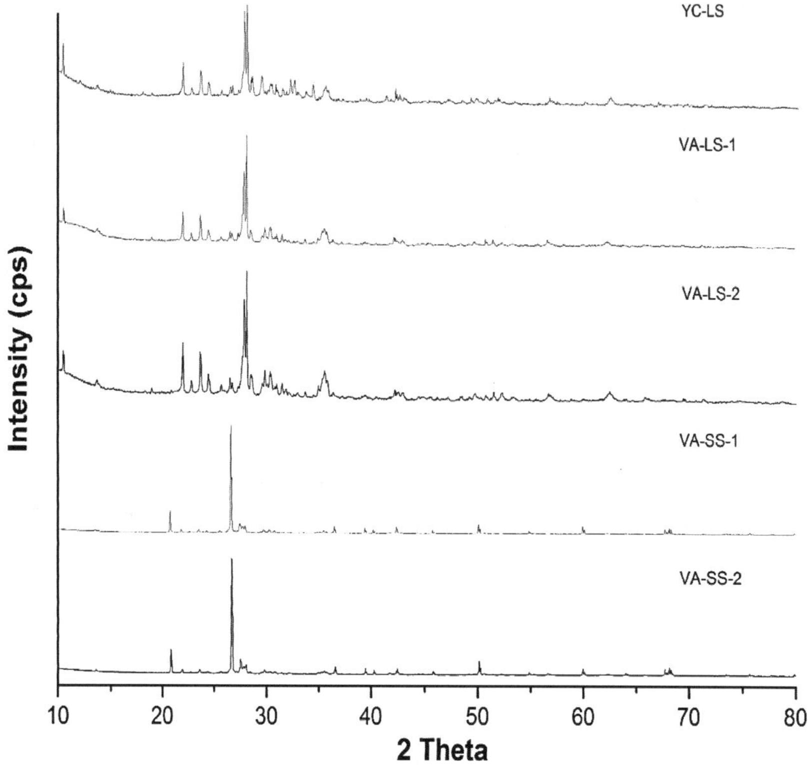

Figure 2. XRD patterns of cement mortar (YC-LS) and volcanic ash based geopolymer mortars (VA-LS-1, VA-LS-2, VA-SS-1, and VA-SS-2)

4.4. Microstructural analysis of mortars

4.4.1. Scanning Electron Microscopy (SEM). Microstructural characteristics of volcanic ash based geopolymer mortars by alkaline activation and different aggregates are shown in figure 3. The images reveal that dissolution of volcanic ash particles and geopolymeric matrix increases with sodium silicate in NaOH solution 12M. Additionally, the images show the presence of micropores and bubbles of water (holes), which influence the compression strength of mortars. According to different studies, it can be inferred that there are areas with unfinished geopolymerization due to the agglomeration of volcanic ash particles which also have influence on the compression strengths of mortars [21], [14] and is reported in this work. Therefore, the high compression strength of VA-SS samples could be related to a better geopolymerization than VA-LS samples.

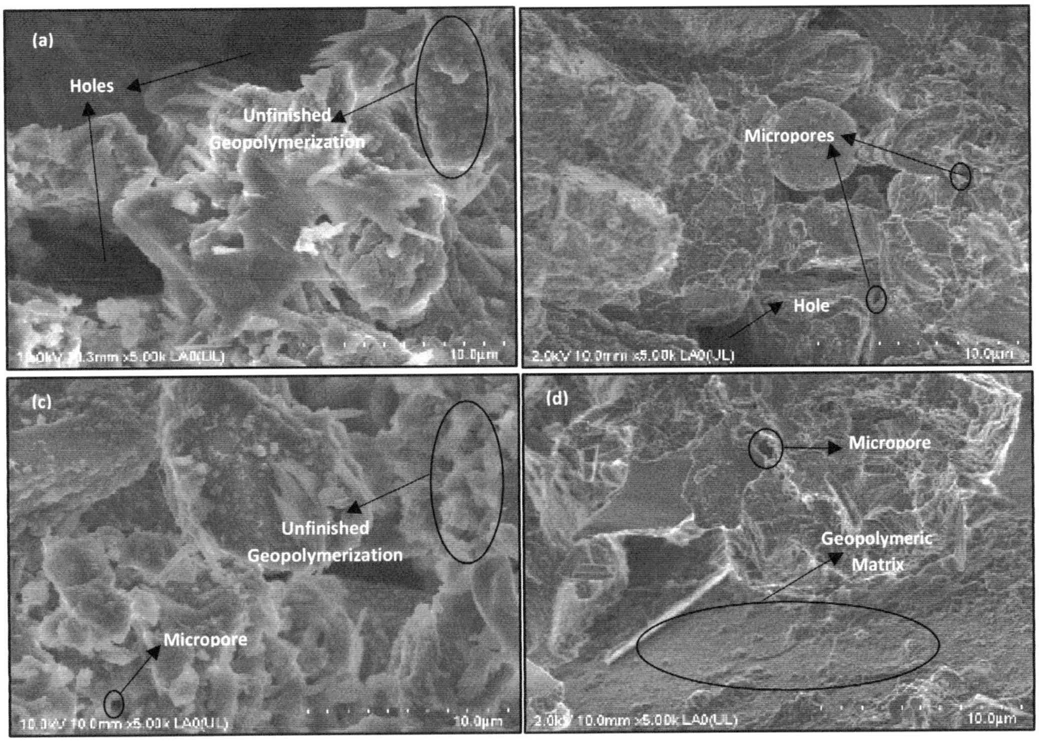

Figure 3. SEM images of volcanic ash based geopolymer mortars (a) VA-LS-1, (b) VA-LS-2, (c) VA-SS-1, and (d) VA-SS-2.

5. Conclusions

The synthesis of volcanic ash based geopolymer mortars was accomplished using different aggregates such as local sand and standard sand with two alkaline solutions, NaOH and Na(OH)+Na_2SiO_3, those initial materials influence on the properties of the geopolymer mortars. The chemical composition analysis of precursor materials confirms the presence of silicate, alumina and calcite, which have no variation on the diffraction patterns of the geopolymer mortars.

The physical properties and the compression strengths for volcanic ash geopolymer mortars are more attractive than cement mortars, allowing the use of volcanic ashes for the production of geopolymers in construction. Samples with NaOH+ Na_2SiO_3 have lower density due to an excessive alkaline solution followed by an unfinished geopolymerization.

According to the gradual decrease on compression strength of volcanic ash based geopolymer mortars, it can be inferred that the aggregates have also influence on the properties due to the different values of samples with LS and SS. It is also considered that the compression strength is related to the presence of calcite, porosity and agglomeration of volcanic ash that could not react to the alkaline solution which can be noticed in the SEM images.

6. References

[1] C. Shi, A. F. Jiménez, and A. Palomo, "New cements for the 21st century: The pursuit of an alternative to Portland cement," *Cem. Concr. Res.*, vol. 41, no. 7, pp. 750–763, 2011, doi: 10.1016/j.cemconres.2011.03.016.

[2] M. Hussain, R. Varely, Y. B. Cheng, Z. Mathys, and G. P. Simon, "Synthesis and thermal behavior of inorganic-organic hybrid geopolymer composites," *J. Appl. Polym. Sci.*, vol. 96, no. 1, pp. 112–121, Apr. 2005, doi: 10.1002/app.21413.

[3] P. Duxson, *Geopolymer precursor design*. Woodhead Publishing Limited, 2009.

[4] D. Bondar, C. J. Lynsdale, N. B. Milestone, N. Hassani, and A. A. Ramezanianpour, "Effect of adding mineral additives to alkali-activated natural pozzolan paste," *Constr. Build. Mater.*, vol. 25, no. 6, pp. 2906–2910, 2011.

[5] O. Macedo, "Investigación sobre volcanes activos en el sur del Perú: Reporte Técnico Especial," 2013.

[6] W. K. Part, M. Ramli, and C. B. Cheah, "An overview on the influence of various factors on the properties of geopolymer concrete derived from industrial by-products," *Constr. Build. Mater.*, vol. 77, pp. 370–395, 2015, doi: 10.1016/j.conbuildmat.2014.12.065.

[7] J. Davidovits, "Geopolymers: inorganic polymeric new materials," *J. Therm. Anal. Calorim.*, vol. 37, no. 8, pp. 1633–1656, 1991.

[8] A. Palomo, A. Fernández-Jiménez, and M. Criado, ""Geopolymers": same basic chemistry, different microstructures," *Mater. Construcción*, vol. 54, no. 275, pp. 77–91, 2004.

[9] J. N. Yankwa Djobo, A. Elimbi, H. Kouamo Tchakouté, and S. Kumar, "Mechanical properties and durability of volcanic ash based geopolymer mortars," *Constr. Build. Mater.*, vol. 124, pp. 606–614, 2016, doi: 10.1016/j.conbuildmat.2016.07.141.

[10] S. Chuah *et al.*, "The properties of fly ash based geopolymer mortars made with dune sand," *Mater. Des.*, vol. 92, pp. 571–578, 2016, doi: 10.1016/j.matdes.2015.12.070.

[11] A. M. Castañón, S. G. Granda, A. M. Guerrero, and F. Gómez-Fernández, "A research of the mineralogy phases of clinker in a Spanish cement using the method of Rietveld," *DYNA*, vol. 79, no. 173 PART I, pp. 41–47, 2012.

[12] R. E. C. Elisa V. Pannunzio-Miner, Mireya del C. Villegas Sanzana, "Análisis Cuantitativo de las Fases Presentes en Cemento Portland y Clinker por Medio de Análisis Rietveld de Patrones de Difracción de Rayos X de Polvos.," *Matéria*, vol. 8, pp. 270–276, 2003, [Online]. Available: http://www.materia.coppe.ufrj.br/sarra/artigos/artigo10272%0A%0A.

[13] H. K. Tchakoute, A. Elimbi, E. Yanne, and C. N. Djangang, "Utilization of volcanic ashes for the production of geopolymers cured at ambient temperature," *Cem. Concr. Compos.*, vol. 38, pp. 75–81, 2013.

[14] P. N. Lemougna, U. F. Chinje Melo, M. P. Delplancke, and H. Rahier, "Influence of the chemical and mineralogical composition on the reactivity of volcanic ashes during alkali activation," *Ceram. Int.*, vol. 40, no. 1 PART A, pp. 811–820, 2014, doi: 10.1016/j.ceramint.2013.06.072.

[15] B. V. Bahoria, D. K. Parbat, and P. B. Nagarnaik, "XRD Analysis of Natural sand, Quarry dust, waste plastic (ldpe) to be used as a fine aggregate in concrete," *Mater. Today Proc.*, vol. 5, no. 1, pp. 1432–1438, 2018, doi: 10.1016/j.matpr.2017.11.230.

[16] L. Gutiérrez De López, "El Concreto Y Otros Materiales Para La Construcción," *2003*, pp. 1–29, 2003.

[17] R. Siddique, "Properties of concrete made with volcanic ash," *Resour. Conserv. Recycl.*, vol. 66, pp. 40–44, 2012, doi: 10.1016/j.resconrec.2012.06.010.

[18] W. K. W. Lee and J. S. J. Van Deventer, "The interface between natural siliceous aggregates and geopolymers," *Cem. Concr. Res.*, vol. 34, no. 2, pp. 195–206, 2004, doi: 10.1016/S0008-8846(03)00250-3.

[19] J. Temuujin, A. Van Riessen, and K. J. D. MacKenzie, "Preparation and characterisation of fly ash based geopolymer mortars," *Constr. Build. Mater.*, vol. 24, no. 10, pp. 1906–1910, 2010, doi: 10.1016/j.conbuildmat.2010.04.012.

[20] J. Baenla, J. B. Bike Mbah, I. B. Djon Li Ndjock, and A. Elimbi, "Partial replacement of low reactive volcanic ash by cassava peel ash in the synthesis of volcanic ash based geopolymer," *Constr. Build. Mater.*, vol. 227, p. 116689, 2019, doi: 10.1016/j.conbuildmat.2019.116689.

[21] A. De Rossi, M. J. Ribeiro, J. A. Labrincha, R. M. Novais, D. Hotza, and R. F. P. M. Moreira, "Effect of the particle size range of construction and demolition waste on the fresh and hardened-state properties of fly ash-based geopolymer mortars with total replacement of sand," *Process Saf. Environ. Prot.*, vol. 129, pp. 130–137, 2019, doi: 10.1016/j.psep.2019.06.026.

Acknowledgments
The authors are grateful to the Materials Science and Engineering and Chemical Engineering Laboratory from Carlos III University of Madrid.

Application of Glass and Fan Shells to a Clay Soil to Increase its Mechanical Properties

Heiner Lopez Jara[1], **Brandon Bravo Barrionuevo**[2] **and Carlos Fernández Díaz**[3*]

1 Faculty of Civil Engineering, Peruvian University of Applied Sciences, Av. Gral. Salaverry 2255, San Isidro, Lima, Perú.
2 Faculty of Civil Engineering, Peruvian University of Applied Sciences, Av. Gral. Salaverry 2255, San Isidro, Lima, Perú.
3 Faculty of Civil Engineering, Peruvian University of Applied Sciences, Av. Gral. Salaverry 2255, San Isidro, Lima, Perú.

*carlos.fernandez@upc.edu.pe

Abstract. Improving the mechanical properties of a clayey soil is one of the best options to avoid future structural failures in buildings and is cheaper than replacing all the material. Therefore, this article proposes the use of recycled glass and fan shells as reinforcement materials. This article presents an experimental study to evaluate the mechanical properties of a pure and mixed soil. The clay soil was mixed with 7% of glass (PV) and with 3%, 6%, 10%, 12% and 15% of Fan Shells (PCA) duly crushed and passed through sieve #100. Tests of sieve granulometry, sedimentation granulometry, Atterberg limits, modified proctor and drained consolidated direct cut were performed. This allowed comparing all the data obtained and defining the optimal percentage of the mixture in which the clay improves its mechanical properties. According to the tests carried out, the proportion that has 7% glass and 6% Fan Shells has better results because there is an improvement in its dry density from 1,784 g / cm3 to 1,847 g / cm3, its moisture content increases from 9.4% to 12.1%. In addition, its friction angle improves from 28.9 ° to 32 ° and cohesion from 0.05 kg / cm2 to 0.1 kg / cm2. These results verify that the properties of the clay soil are improved.

1. Introduction

Clayey soils turn out to be one of the problems with the highest incidence in construction, because the increases in volume and the mechanical properties of this land where it is built are not announced in a uniform way, on the contrary, they cause growth in different areas and at the moment of contracting they produce settlements, which can severely damage the structures.

In Peru there are several areas with clay soil problems, and we refer to the province of Talara that presents large tracts of land with the presence of this soil and this adds to the lack of alternatives to control the expansive behaviour and low resistance of the mechanical properties that this soil presents in buildings.

Building on these clayey soils, which by their nature have a low bearing capacity, means that the requirements necessary to carry out civil works projects are not met. This implies an increase in costs to try to stabilize these problematic soils with different alternatives.

Geotechnical engineering offers solutions for various problems that occur with clayey soils such as pile construction and soil improvement. In the improvement alternative, there are several

investigations that present the technique of stabilization and improvement of mechanical properties with the application of mollusk valves and crushed glass, as evidenced by their results:
D. Koteswara Rao [1] did his research in India testing a soil mixture with mollusk valves and ferric chloride. The proportion of the mollusc shells was from 0% to 15% and the Ferric Chloride was from 0.5% to 2.5%. Atterberg limits, specific gravity, compaction and CBR tests were performed. The results showed improvements in the mechanical behaviour of the clay soil with 10% of mollusk valves and 1.5% of ferric chloride.

Maheshwari G [2] carried out an investigation evaluating the effect of mollusk shells that are naturally available materials on the coasts of the sea and bitumen emulsions that act as a binder between the clayey soil particles, which prevent the entry of water into the soil. Uniaxial compression and CBR tests were performed. With a 16% addition of mollusk shell powder, it significantly reduces its plasticity indexes and the CBR values of the soil samples increased considerably.

Al-Hassnawi [3], carried out a study with the intention of improving the plasticity, support, consolidation and cut resistance of clay when interacting with glass. The residual glass was used as an additive material that passes through the # 200 sieve in percentages of 3, 5, 7 and 9%. Tests of consistency limits, unconfined compression and CBR were performed. It was concluded that the additive had a positive impact on the soil and had improved properties. The liquid limit, the plastic limit and the plasticity index decreased with a 7% increase in glass powder.

H. Canakci [4], did a study with the intention of observing improvements of the clayey soil due to the addition of residues of glass powder of soda and lime (WSLGP). The soda lime glasses were crushed and sieved through the # 200 mesh (75 μm) and were also mixed with clay at 3%, 6%, 9% and 12%. The strength and consistency test were carried out on the mixed samples after curing. The test results indicated that the addition of 12% WSLGP to the clay has a significant effect on the strength and consistency properties of the clay.

In all these previous investigations positive results were seen such as the improvement of the plasticity index, the capacity to support the load, an increase in maximum dry density and an increase in CBR. In order to solve this problem, in this research, stabilization using two external agents such as mollusk shells (PCA) and crushed glass (PV) is proposed as an alternative for improving the mechanical properties of clay soil.

2. Experimental program

2.1. Materials

2.1.1. Clay *soil.* The clay soil used to improve its mechanical properties is found in the town of Piura, Peru; its coordinates are: 4 ° 35'19.0 "S 81 ° 16'15.3" W. The physical properties of the clay soil were identified and investigated according to technical standards (ASTM). The tests carried out were: sieve granulometry, sedimentation granulometry and consistency limits.

The soil studied is made up of 98.1% fines and 1.9% sand. This clay has a CL classification (low plasticity sandy clay) according to SUCS and for AASHTO A-7-6 (26) [5]. Table 1 shows the characteristics.

Table 1- Characteristics of natural soil

Soil Properties	Results
Consistency limits	
Liquid Limit (LL)	47%
Plastic Limit (LP)	23%
Plasticity Index (IP)	24%

SUCS classification	CL
AASHTO classification	A-7-6(26)
Soil composition	
Gravel	0.00%
sand	1.90%
fine particles	98.10%

2.1.2. Fan shells. The scallops used for this research were extracted from the fishing dumps in the area. This element was used, since one of its predominant chemical components is calcium carbonate (CaCO3) [6], these are the same minerals that Cal contains. See figure 1.

Figure 1. Fan shells.

2.1.3. Glass (recycled bottles*).* The glass used was collected from junkyards in the area. This material was used because it contains calcium carbonate (CaCO3) and Silica sand (SiO2) [7], elements that work well with fine soils. See figure 2.

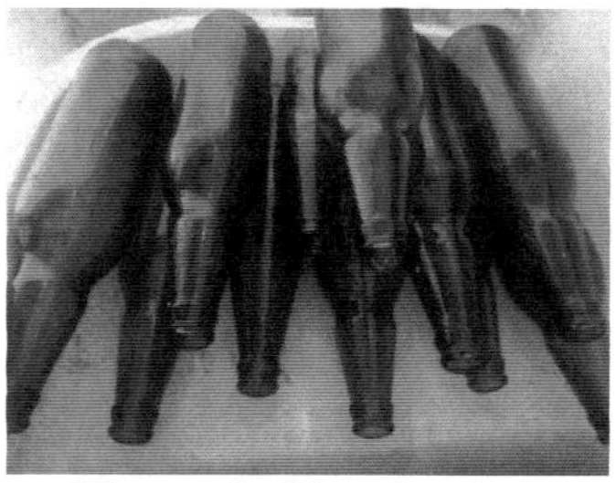

Figure 2. Glass bottles.

2.2. Preparation of mixtures

For the preparation of the mixtures, references from previous academic articles and theses were taken. This allowed defining the sample percentages for each test. The particle size with which we worked for the mixtures of PV (glass powder) and PCA (fan shell powder) was retained on the N°100 sieve, this size was used because as the fine soil is granular, it can better grip the particles and thus have better soil stabilization. A nomenclature was assigned for each mix. See table 2.

Table 2. Mix Design.

Mixture	% Soil	% Mollusk Shells	% Glass Powder	Symbology
Natural Soil	100	0	0	NS
Mixture 1	90	3	7	M1
Mixture 2	87	6	7	M2
Mixture 3	83	10	7	M3
Mixture 4	81	12	7	M4
Mixture 5	78	15	7	M5

2.3. Laboratory tests

The tests were carried out for the mixtures defined in table 2 according to the ASTM standard, which were:

2.3.1. Atterberg limits test. This test allows determining the different states of the soil such as liquid limit (LL), plastic limit (LP) and plasticity index (IP). With the mixtures already defined in table 2, the tests were carried out according to ASTM D4318 [8].

2.3.2. Modified Proctor Test. It aims to achieve maximum dry density for optimum humidity in order to improve the soil [8]. Being a repetitive compaction process, the objective is to reduce the volume of voids, generating an increase in density. With the mixtures already defined in table 2, the tests according to ASTM D-1557 were carried out.

2.3.3. Drained Consolidated Direct Shear Test. This test determines the resistance to the shear stress of the soil and determines the resistance parameters such as cohesion and friction angle [8]. The test was carried out applying a normal force to the ground of 0.5 kg/cm2 (49 kPa), 1.0 kg/cm2 (91.1 KPa) and 2.0 kg/cm2 (196.1 KPa). The vertical movement is read by placing a deformmeter on top. The applied deformation speed was 0.00492 in / min, since it must be slow to guarantee drainage conditions equivalent to zero pressure. Compaction was carried out with the humidity and unit weight conditions obtained from the control test. In addition, the specimens were allowed to cure for approximately one day.

3. Analysis and results

A comparative analysis of the results obtained in the Atterberg limit tests, modified proctor and drained consolidated direct cut was performed on the natural soil and on the mixtures defined in table 2.

3.1. *Atterberg* limits test

Carrying out the tests, figure 3 was obtained, where the results of the test for natural clay and the mixtures are summarized.

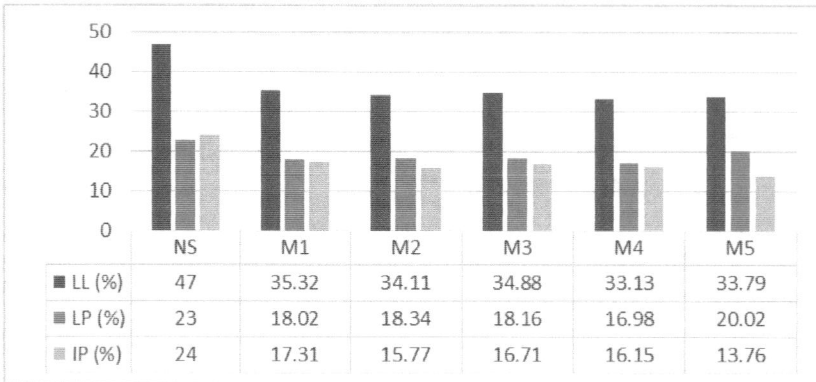

Figure 3. Atterberg limits tests.

It is evident that natural clay has a higher liquid limit and tends to decrease when adding the percentages already defined in table 2. This is generated, since the additive used for the study has a particle size of 150 μm, all that is retained in mesh # 200. Furthermore, by reducing the percentage of water (LL), a reduction in the volumetric variation of the soil is generated.

3.2 Modified proctor test
From the tests performed, a summary of all modified Proctor tests was developed. See figure 4.

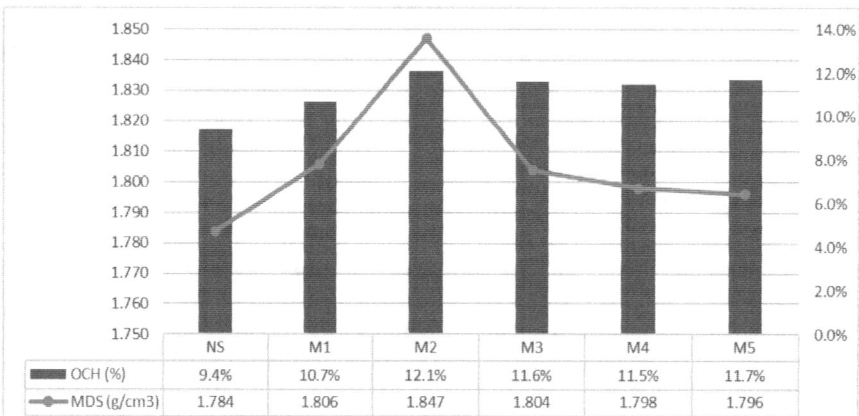

Figure 4. Summary of modified proctor results.

It is observed that the maximum dry density increases compared to natural clay. This has a maximum dry density (MDS) of 1,784 kg / cm3 and a maximum moisture content (OCH) of 9.4%. When the percentage is increased to 6% PCA-7% PV (M2) there is a peak increase in MDS of 1,847 kg / cm3 and an OCH of 12.1%. When 10% PCA-7% PV (M3) is added, it begins to have a decreasing trend in MDS and OCH.

3.3 Direct cut test consolidation drained
In the direct shear tests, the Tangential Deformation vs. Shear Stress graphs were performed at different normal stresses already defined in the experimental program. See figure 5, 6 and 7. This will allow us to identify the maximum shear stresses and find the parameters (see table 3) of friction and cohesion angle.

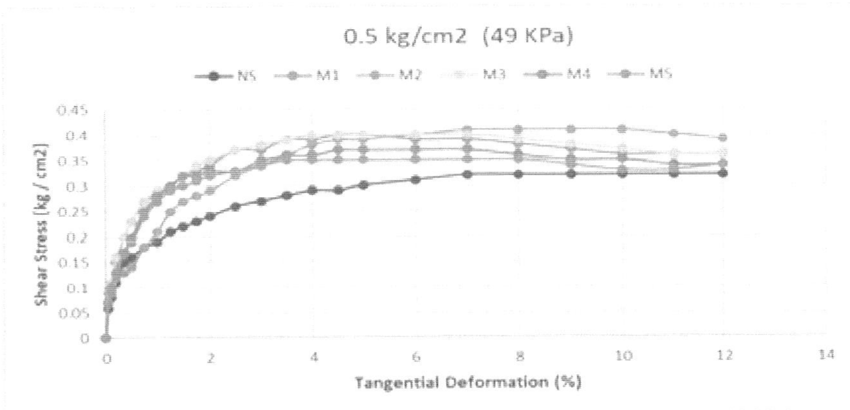

Figure 5. Tangential Deformation (%) vs Shear Stress (kg / cm2) of all the mixtures at 0.5 kg / cm2.

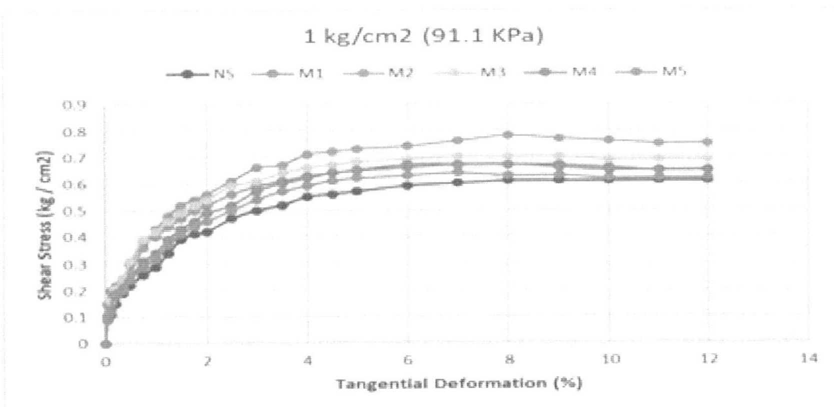

Figure 6. Tangential Deformation (%) vs Shear Stress (kg / cm2) of all the mixtures at 1 kg / cm2.

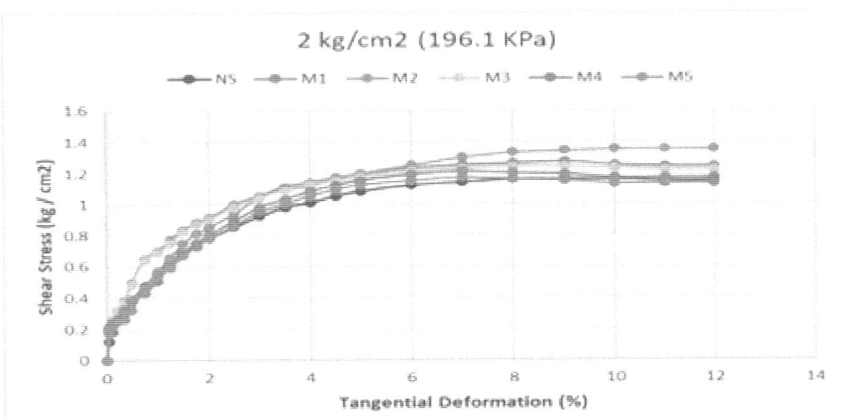

Figure 7. Tangential Deformation (%) vs Shear Stress (kg / cm2) of all the mixtures at 2 kg / cm2.

Table 3. Maximum shear stress.

Normal effort (kg / cm2)	Maximum Shear Stress (kg / cm2)					
	NS	M1	M2	M3	M4	M5
0.5	0.32	0.4	0.41	0.4	0.37	0.35
1	0.61	0.67	0.78	0.7	0.67	0.64
2	1.16	1.27	1.35	1.24	1.21	1.17

Identifying the maximum shear stresses, the failure envelope of each sample was plotted with the Mohr circle. See figure 8 and table 4.

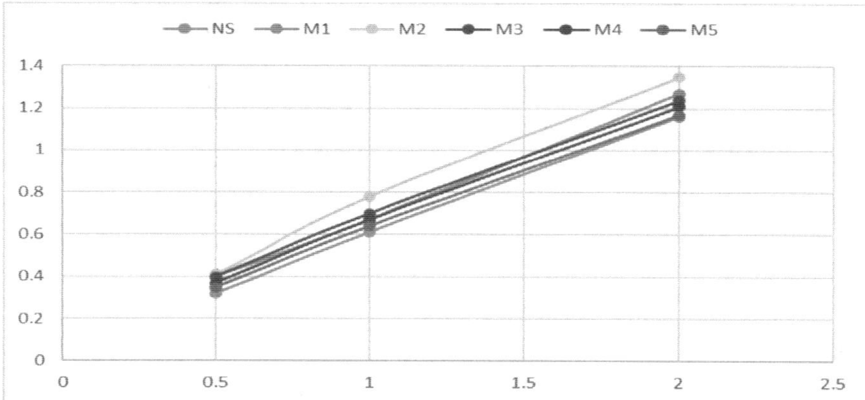

Figure 8. Normal Effort vs. Cut Effort of all mixes.

Table 4. Summary of resistance parameters.

	NS	M1	M2	M3	M4	M5
Ø	28.9	30.5	32	29.8	28.6	28.3
Cohesion	0.05	0.07	0.1	0.08	0.08	0.06

It is evident that there is a trend of increasing resistance parameters in the design mixtures, but then it begins to decrease with increasing percentages of PV and PCA. The M2 mix is the one that has the best results, since the natural clay soil has a friction angle of 28.9 ° and a cohesion of 0.05 kg / cm2 and increases in M2 its friction angle to 32 ° and its cohesion to 0.1 kg / cm2.

4. Conclusions

• The liquid limit of the natural soil is 47% and when adding the first mixture, M1 decreases to 35.32% and this tendency to decrease for the other mixtures continues.

• In DMS, the highest result obtained from all the mixtures was M2, since it increases by 0.063 g / cm3 with respect to natural soil (NS). This occurs because the scallop behaves as a binder.

• In the consolidated direct cut drained it is observed that the M2 mix has better results. When the applied normal tension is 0.5 kg / cm2 (49 KPa), the shear resistance of the natural soil compared to the M2 mixture increases from 0.32 kg / cm2 to 0.41 kg / cm2, in the same way for the applied normal tension of 1 kg / cm2 (91.1 KPa) the cut resistance increases from 0.61 kg / cm2 to 0.78 kg / cm2, and when the applied tension is 2 kg / cm2 (196.1 KPa) the cut resistance increases from 1.16 kg / cm2 to 1.35 kg / cm2.

• The friction angle of M2 increased by 10.7% with respect to the natural NS soil and its cohesion increased by 100%, going from 0.05 kg / cm2 to 0.1 kg / cm2.
• The use of fan shells and glass contribute to the environment, since they are recycled elements. In addition, there is an incentive to discover new stabilization methods for clay soils.
• With this research it was shown that the stabilizing elements for clay soil achieve a significant increase in mechanical properties, thus achieving an improvement in the soil for constructions.

5. References

[1] Kotcswara D and Vineela S 2017 A laboratory study on the performance of expansive soil subgrade treated with seasheel powder and ferric chloride, *International Journal of Technical Innovation in Modern. Engineering & Science.* **4** 1040-1045
[2] Maheshwari G B and Ravikumar B 2015 Study on Stabilization of Soil Using Sea Shell and Bitumen Emulsion, *International Journal of Innovative Research in Science Engineering and Technology.* **4** 5875-5882
[3] Mahdi Z and Alhassnawi N 2018 Assessment of subgrade soil improvement by waste glass powder, *International Journal of Civil Engineering and Technology* **9** 12-21.
[4] Canakcia Hanifi, ALKakia Aram and Celika Fatih 2016 Stabilization of Clay with Waste Soda Lime Glass Powder, *Elsevier Ltd* **161** 600-605
[5] Braja M D 2012 Foundations Engineering Fundamentals *CENGAGE Learning* (Mexico: International Thomson) 44-52.
[6] FONDEPES 2018 *Manual de conchas de Abanico* (Lima: Ministerio de la Producción) pp 15 – 18
[7] Rincón J M 2006 *Materias Primas para la Industria del Vidrio* Laboratorio de Materiales Vítreos y Cremaicos (España: Instituto E. Torroja de Ciencias de la Construcción) 50-73
[8] Pontificia Universidada Católica del Perú 2012 *Guía de Mecánica de Suelos* (Lima: Pontificia Universidada Católica del Perú) 2-106

IOP Conf. Series: Materials Science and Engineering 1054 (2021) 012005 doi:10.1088/1757-899X/1054/1/012005

Use of Textile Waste as an Addition in the *elaboration of an Ecological* Concrete Block

J Anglade[1*], E Benavente[1], J Rodríguez[2], A Hinostroza[3]

1 Bachelor, Civil Engineering Program, Universidad Peruana de Ciencias Aplicadas, Lima, Perú
2 Full Professor, Civil Engineering Program, Universidad Peruana de Ciencias Aplicadas, Lima, Perú
3 Professor, Civil Engineering Program, Universidad Peruana de Ciencias Aplicadas, Lima, Perú

*U201518908@upc.edu.pe

Abstract. The textile industry has grown significantly in recent years, reaching a global fiber production of 53 million tons which 12 % are recycled; Construction sector has been using more and more recycled materials from different industrial sources, to apply them in their constructions and to reduce CO_2 emissions and final energy consumption. The present study aims to study the behavior of concrete blocks of $f'_c = 210$ kg/cm^2 adding polyester textile waste with 3 %, 6 %, 9 %, 12 % and 15 %; void content, compressive strength and thermal conductivity decrease, and water absorption, acoustic insulation and unit price increase by 3 %, 34 % and 16 % compared to conventional concrete block.

1. Introduction

The accelerated growth of the textile industry has caused problems for environment, in 2015 the emissions of Greenhouse Gases (GHG) totaled 1200 million tons of CO_2 equivalent, this due to the fact that in its production per 1 Tn of textile is produced 17 Tn of CO_2 [1]. Similarly, its textile waste, whose level of contamination varies depending on the type of fiber used, originate methane gas and ammonia [2], [3]. There are some initiatives within the circular economy to minimize this waste, which are oriented towards its reuse in the textile industry or in other sectors such as construction [4], which is considered one of the most polluting due to its high level of CO_2 emissions [5], [6], being buildings responsible for the third part [7], [8].

In order to reduce these emissions, materials can be used in their construction that reduce the energy consumption used in the exploitation of the aggregates as natural resources [9], and that making use of traditional techniques updated with locally available materials [10] , improve their thermal conductivity and are more insulating; In this line, we can propose the need to change the composition of the 0.12 m and 0.15 m conventional concrete blocks used in the walls of the buildings, because with these an average thermal insulation of 0.159 m^{2}K/W and 0.171 m^{2}K/W is achieved [11]. In this regard, [12] carried out an investigation in a masonry system composed of concrete blocks coated with cement mortar and brick coated, also with cement, both materials separated by an air chamber, to compare the thermal conductivity with 2 types of 100 % acrylic fabric; first, with woven fabric residues and the other, with woven fabric sub-residues, finding a higher thermal conductivity for the mortar equal to 1.3 W/m°C, and lower values for the woven fabric and woven fabric with sub-residues, these being 0.044 W/m°C and 0.103 W/m°C respectively; these values coincide with investigations carried out by [13], [14]; In this sense, an alternative for improvement would be to elaborate a concrete block with polyester textile residue. Polyester is a synthetic polymer, whose main source of raw material is petroleum, it can be used in the form of fiber or filament, it is strong and dimensionally stable; it has high resistance to humidity and biotic attacks, it does not contain insecticides or glues [15]. It has good thermal and acoustic insulation, its resistance to fire is good and it is less degradable in humid conditions [15], [16], and it is used in industrial and technical applications such as woven and non-woven fabrics [17].

In this regard, we can mention some referenced investigations on the characterization of textile waste: Water absorption is measured in [18] for concretes with additions of 0.5 %, 1.0 % and 2.0 % of carpet, linen and sisal fibers; finding that it is 4.3 %, 7.1 % and 7.8 % higher than the control sample. [19] found the same trend when studying lightweight concrete blocks with banana fiber additions of 2.5 %, 5.0 % and 7.5 % for different amounts of cement, finding that the water absorption at 28 days was maximum for 7.5 % and equals 201.6 % with respect to the standard block. Regarding the voids content and cost, we can indicate that similar investigations have not been found, but we can mention [20] who finds that the lightweight concrete blocks of 14 cm x 39 cm x 19 cm with the addition of polystyrene beads are 8 % more expensive than conventional concrete blocks, and this is due to the acquisition of the added material.

Regarding the compressive strength, [3] states that the compressive strength at 7 days of concrete blocks with a 25 % replacement of fabric increases by 19 %; and for the age of 28 days, [21], [18] indicate that the increase is between 12.5 % - 21.0 % when varying the percentage of cloth (linen) from 0.5 % - 40 %, and then decreases from 40 % - 94 % with respect to standard concrete. Regarding acoustic insulation, [22] tested fabrics with recycled polyester waste and wool waste, finding that the acoustic coefficient increases with the increase in the frequency range of 50 - 5700 Hz, with the acoustic coefficient being lower in 13.1 % for waste with 100 % recycled polyester fiber compared to waste with 100 % wool; on the other hand, [15] studied the fibers of polyester waste for different mass densities, indicating that the acoustic absorption coefficient increases with increasing mass from 65 - 95 K/gm^3 for frequencies of 100 - 5000 Hz.

With regard to thermal conductivity, [22] after testing fabrics with recycled polyester waste and wool waste, found that the thermal conductivity of waste with 100 % recycled polyester fiber is 9.4 % more than waste with 100 % woollen; Similarly, [15], [23] study fibers of polyester waste for different mass densities, indicating that the thermal conductivity decreases from 13.7 - 29.3 % when the mass of the fabric increases from 65 - 396 Kg/m^3.

This research addresses the low thermal conductivity of conventional concrete blocks in building walls. For this, the percentage of water absorption, the percentage of voids, the cost of manufacturing, the resistance to compression, and the acoustic and thermal insulation and unit cost of the concrete block with polyester have been studied.

2. Materials and methods

2.1. Materials

Portland cement Type I was used according [24]; fine and coarse natural aggregates with a maximum size of Ø ¾ " comply with [25]; Polyester Fabric (PF) was used in sizes of 30 mm x 60 mm and thickness of 2mm, whose characteristics are: density = 0.16 g/cm^3, elongation = 50 - 70, colour = variety; and image are shown in figure 1, drinking water was used to prepare the mixture according [26].

2.2. Methods

2.2.1. Laboratory testing

The mixtures of the concrete blocks with PF were designed for an f'c = 210 Kg/cm^2 with the method [27], Table 1 and Table 2 show the design of mixtures for a Block with Polyester Fabric (BPF), and the number of blocks, ages and test standards; Blocks were made manually in molds measuring 14 cm wide x 19 cm high x 39 cm long, and compacted on a vibrating table according [27]. Their storage was carried out in a closed environment exposed to room temperature for 24 hours according to [28], then they were cured in water following [28] for 7 and 28 days until the date of the tests. For the case of water absorption and voids content, 3 concrete blocks were used per test with 6 % and 12 % TC and were tested after 56 days according [29], [30].

Table 1. Mix design of concrete blocks.

Materials	BPF-0 (Pattern)	BPF-3 (3%PF)	BPF-6 (6%PF)	BPF-9 (9%PF)	BPF-12 (12%PF)	BPF-15 (15%PF)
Proportion (C:F:C)	1:2:3	1:2:3	1:2:3	1:2:3	1:2:3	1:2:3
Cement (Kg)	2.05	1.99	1.93	1.87	1.80	1.74
Fine Aggregate (Kg)	4.592	4.582	4.574	4.565	4.555	4.546
Coarse Aggregate (Kg)	7.319	7.303	7.291	7.276	7.261	7.245
Water (Kg)	1.44	1.39	1.35	1.31	1.26	1.22
Polyester Fabric (Kg)	0.00	0.25	0.49	0.74	0.98	0.12

Table 2. Number of blocks, ages and standards according to tests.

Tests type	BPF-0 28 days	BPF-3 28 days	BPF-6 28 days	BPF-9 28 days	BPF-12 28 days	BPF-15 28 days	Methods
Compressive Strength	3	3	3	3	3	3	[31]
Acoustic isolation	2	2	2	2	2	2	[32]
Thermal isolation	2	2	2	2	2	2	[33]

2.2.2. Unit price

The analysis of unit prices of the conventional concrete block and concrete block with fabric was made with the studies of [34], [35], having considered the price of polyester as the sale price in Lima of S/. 0.50 / Kg and the dimensions of the polyester of 3 cm x 6 cm.

3. Fresh concrete

3.1. Water absorption

Figure 2 shows the influence of PF on the water absorption of the concrete block, the behavior increases when moving from BPF - 6 block to BPF - 12 block, reaching the values of 3.93 % and 4.03 % respectively, which represents a 2.48 % increase in absorption with respect to the BPF - 6. [36] studied the effect of PF on the water absorption of the concrete block for the age of 56 days, finding that there is an increase when going from 20 gr to 60 gr, reaching the values of 14.01 % and 18.76 % respectively, which represents a 25.27 % increase in absorption compared to block with 20 gr. [37] evaluated the incidence of the cotton fabric fiber on the water absorption of the limestone block for the age of 28 days, finding that there is an increase when going from 32 gr to 130 gr, reaching the values of 12.6 % and 27.2 %, which represents an increase of 53.7 % of absorption with respect to block of 32 gr. [38] indicates that this behavior is due to the presence of a more porous structure in the concrete.

3.2. Void content

Figure 3 shows the influence of PF on the void content in the concrete block, the behavior decreases when moving from the BPF - 0 block to the BPF - 12 block, reaching the values of 42 % and 39.6 %, which represents a 5.71 % decrease in absorption in relation to BPF - 0. [39] studied the effect of the fly ash on the void content of a polymeric concrete block for the age of 28 days, finding a decrease from 6 % to 9 % of addition, reaching values of 22 % and 13 % respectively, which represents a lower void content of 40.91 % compared to the 6 % block. [40] evaluated the incidence of the density of the polypropylene fiber on the void content of a permeable concrete block for the age of 28 days, finding a decrease when going from a density of 1710.23 kg/m³ to 1837.92 kg/m³, reaching values of 32.38 % and 25.27 % respectively, which represents a lower void content of 28.14 % in relation to block with 1710.23 kg/m³. [41] indicates that this behavior may be due to the filling of the concrete voids.

Figure 1. PF sample.

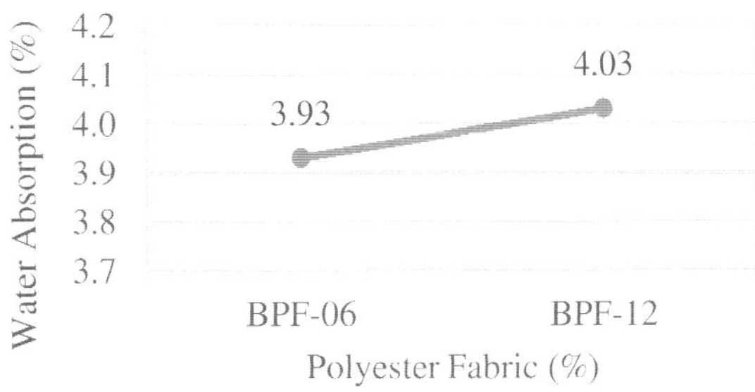

Figure 2: Water absorption.

4. Hardened concrete

4.1. Compressive strength

Figure 4 shows the influence of PF on the compressive strength of the concrete block at the age of 28 days. Resistance decreases, reaching the value of 24.23 Kg/cm^2 for the BPF - 15 block, which represents 49 % less than BPF - 0 block (standard). [42] studied the effect of the percentage of fabric on the compressive strength of the concrete block at the age of 28 days for two mixtures with 1 % and 2 % addition of 4 types of fabrics: polyester, cotton, silk and rayon, and block dimensions of 40 mm x 40 mm x 160 mm, finding a lower resistance value of 183.96 Kg/cm² for 2 %, which represents 43.09 % less with respect to the 0 % block (standard). [43] evaluated the incidence of the percentage of polyester fabric on the compressive strength of the concrete block at the age of 28 days, for mixtures with 0.2 %, 0.4 %, 0.6 %, 0.8 %, 1 %, 1.2 %, 1.4 % and 1.6 %, and block dimensions of 150 mm x 150 mm x 150 mm, finding that the resistance decreases, reaching a lower value of 183.96 Kg/cm² for 1.2 %, which represents 34.25 % less with respect to the 0.2 % block (standard). [44] indicates that this behavior may be because the materials with polymeric origin resist less to compression forces, their deformation is low and they present a weak bond between them compared to fine aggregates; which produces cracks around it and the existence of voids inside the concrete.

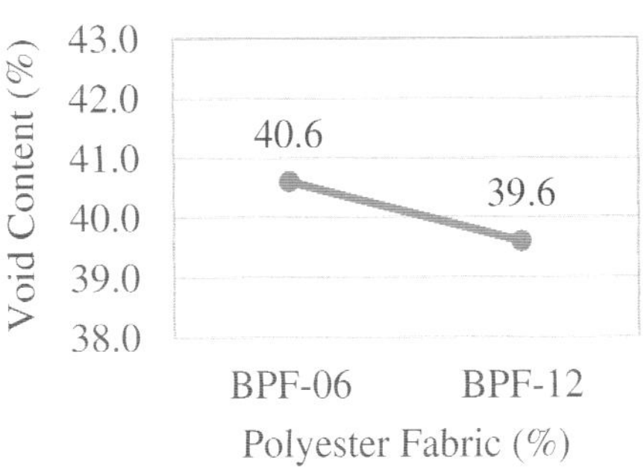

Figure 3. Void content.

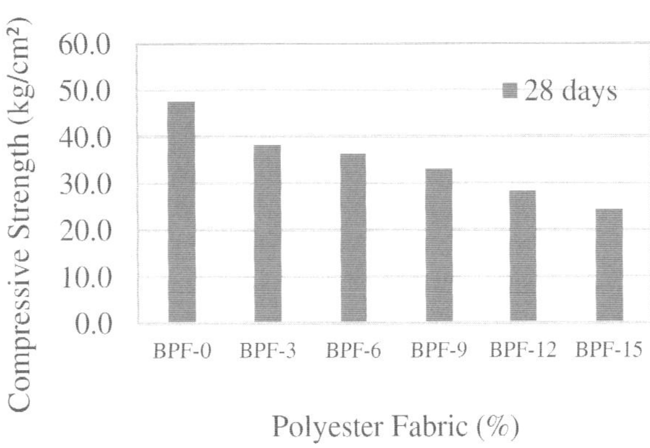

Figure 4. Influence of PF on the compressive strength.

4.2. Acoustic isolation

Figure 5 shows the influence of PF on the sound pressure difference of the concrete block at the age of 28 days. The sound pressure difference increases, reaching the value of 34.76 dB for BPF - 15 block, which represents 34.3 % more than BPF - 0 block (standard). [45] studied the effect of the amount of polyester fabric on the sound pressure difference in a 3.55 m x 2.85 m x 0.06 m drywall plate at 28 days, for mixtures with 20 % and 35 %, finding that the sound pressure difference increases, reaching the highest value for 35 %, this being 43 dB, which represents 2.4 % more with respect to the standard amount of 20 g/m² (standard). [46] evaluated the incidence of polyester fabric on the sound pressure level, for samples of 21.0 kg/m³, 34.3 kg/m³ and 65.5 kg/m³ of 400 mm x 200 mm x 18 mm polyester, finding that the pressure level noise decreases with the increase in the amount of polyester, obtaining the lowest value for the sample of 65.5 kg/m³, this being 56 dB(A), which represents 12.5 % less compared to the sample of 21.0 kg/m³; which means that better insulation is thus achieved. [47] indicates that this behavior is due to greater friction between the fibers and the air.

4.3. Thermal conductivity

Figure 6 shows the influence of PF on the coefficient of thermal conductivity of the concrete block at the age of 28 days. In it, it is seen that the coefficient decreases, reaching the value of 1.54 for BPF - 15 mixture, which represents 5.5 % less than the BPF - 0 block (standard). [46] studied the effect of the fabric on the coefficient of thermal conductivity at 28 days for 2 samples with 100 % polyester and 95 % polyester plus 5 % addition lycra on a cylindrical sample with dimensions of 10 cm x 12 cm in diameter, finding that there is a decrease in the coefficient of thermal conductivity when the percentage of fabric varies, reaching the lowest value for the sample with 100 % being 0.0543 W/mK, which represents 11.05 % less than the sample with 95 % polyester. [48] evaluated the incidence of the percentage of polyester fabric on thermal conductivity in fabric samples of 200 mm x 200 mm x variable widths of 5 %, 15 %, 20 %, 25 % and 30 % with respect to the length, finding that the Thermal conductivity coefficient decreases as the percentage of polyester varies, reaching the lowest value for the sample of 30 %, this being 36.3 W/mK, which represents 1.65 % less than the standard sample with a width of 5 %. [49] indicates that this behavior is due to the strong interaction that exists in heat transfer between fibers and air; likewise [50] states that the amount of air trapped in the matrix also reduces the thermal conductivity.

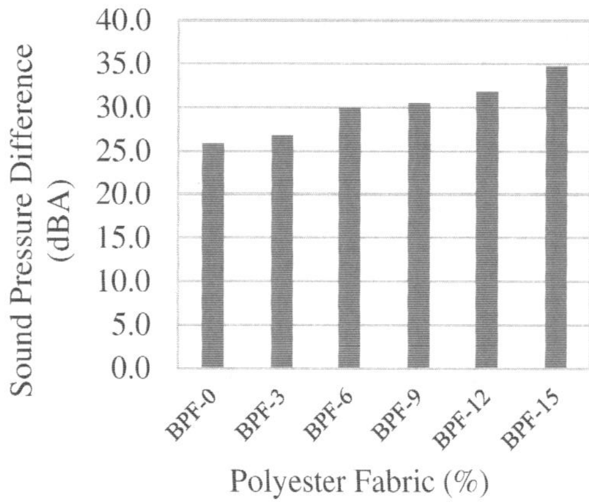

Figure 5. Influence of PF on the sound pressure difference.

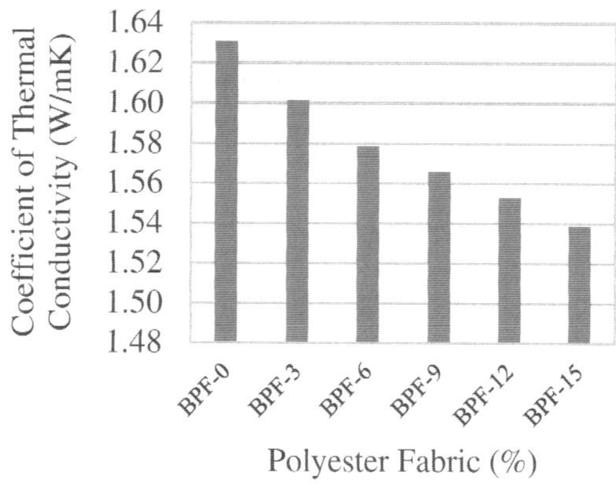

Figure 6. Influence of PF on the coefficient of thermal conductivity.

5. Unit price

Tables 3 and 4 show the analysis of unit prices per m^3 of a conventional concrete block and a concrete block with PF. In them the unit price per m^3 is S/. 3.75 ($ 1.05 USD) for the conventional concrete block and S/. 4.41 ($ 1.21 USD) for block with 15 % polyester, this being 16 % more than the conventional concrete block due to the cost incurred and associated with the production and the added material [51], [52], [53].

Table 3. Analysis of the unit price per m^3 of conventional concrete block.

Performance 120 block/day		Conventional concrete block 39 cm*19cm*14cm			S/.3.75
Description	Unit	Quadrille	Quantity	Price (S/.)	Partial (S/.)
Materials					
Cement	Kg		2.6797	0.42	1.13
Sand	m³		0.0033	56.00	0.18
Gravel	m³		0.0065	45.00	0.29
Water			0.0014	4.82	0.01
					1.61
Workforce					
Laborer	HH	1.0000	0.0667	14.33	0.96
					0.96
Equipment					
Rosa Cometa	HM	1	0.0533	20.00	1.07
Manual tools	%MO		0.0478	2.55	0.12
					1.19

Table 4. Analysis of the unit price per m^3 of concrete block with PF.

Performance 120 block/day		Concrete block with PF 39cm*19cm*14cm			S/.4.41
Description	Unit	Quadrille	Quantity	Price (S/.)	Partial (S/.)
Materials					
Cement	Kg		2.6797	0.42	1.13
Sand	m³		0.0033	56.00	0.18
Gravel	m³		0.0065	45.00	0.29
Water			0.0014	4.82	0.01
Polyester	Kg		0.1200	1.00	0.12
					1.73
Workforce					
Laborer	HH	1.5000	0.1000	14.33	1.43
					1.43
Equipments					
Rosa Cometa	HM	1	0.0533	20.00	1.07
Manual tools	%MO		0.0717	2.55	0.18
					1.25

6. Conclusions

The high thermal conductivity of conventional concrete is due to the fact that, with its current composition, the coefficient of thermal conductivity, the apparent density and the porosity, necessary to manufacture a block with thermal characteristics, are not improved.

Increased water absorption due to the increase in polyester is because fibers needs more water to mix with the concrete.

Void content decreases as the amount of polyester increases because the fibers cover the voids generated in the concrete slab.

Compressive strength decreases because when polyester is added, fibers increase air bubbles and voids inside the concrete slab.

Increase in acoustic insulation is because the apparent density increases and influences the acoustic absorption coefficient, making the concrete block with polyester a better acoustic insulator.

By decreasing the thermal conductivity by increasing the porosity of the concrete, a polyester block is produced that is a better thermal insulator.

Polyester concrete block is 16 % more expensive than conventional concrete block due to the cost for the recycled material and for the payment of personnel who cut the recycled polyester into small dimensions.

7. References

[1] MacArthur E 2017 A new textiles economy: Redesigning fashion's future *Ellen MacArthur Foundation* pp 1–150.
[2] Wang Y 2010 Fiber and textile waste Utilization *Waste and Biomass Valorization* vol 1 pp 135–143.
[3] Jayasinghe I, et al. 2010 Environmental Conservation Efforts in Developing Textile Waste Incorporated Cement Blocks *Tropical Agricultural Research* vol 21 p 126.
[4] Pensupa N, et al. 2017 Recent Trends in Sustainable Textile Waste Recycling Methods: Current Situation and Future Prospects *Topics in Current Chemistry* vol 375 pp 76
[5] Hussin J, Abdul I and Memon A 2013 The Way Forward in Sustainable Construction: Issues and Challenges *Int J of Advances in Applied Sciences* vol 2
[6] Pardo N and Thiel C 2012 Evaluation of several measures to improve the energy efficiency and CO_2 emission in the European single-family houses *Energy and Buildings* vol 49 pp 619–630
[7] Barbero-Barrera M, Pombo O and Navacerrada M 2016 Textile fibre waste bindered with natural hydraulic lime *Composites Part B: Eng* vol 94 pp 26–33
[8] Asdrubali F, D'Alessandro F and Schiavoni S 2015 A review of unconventional sustainable building insulation materials *Sustainable Materials and Technologies* vol 4 pp 1–17
[9] Meng Y, Ling T and Mo K 2018 Recycling of wastes for value-added applications in concrete blocks: An overview *Resources, Conservation and Recycling* vol 138 pp 298-312
[10] Madurwar M, Ralegaonkar R and Mandavgane S 2013 Application of agro-waste for sustainable construction materials: A review *Const and Building Mat* vol 38 pp 872-878
[11] Pérez J, et al. 2011 Estudio Numérico de la Resistencia Térmica en Muros de Bloques de Concreto Hueco con Aislamiento Térmico *Información Tecnológica* vol 22 pp 27–38
[12] Briga-Sá A, et al. 2013 Textile waste as an alternative thermal insulation building material solution *Const and Building Mat* vol 38 pp 155–160
[13] Awal A and Mohammadhosseini H 2016 Green concrete production incorporating waste carpet fiber and palm oil fuel ash *J of Cleaner Production* vol 137 pp 157–166
[14] Zhan B and Poon C 2015 Study on feasibility of reutilizing textile effluent sludge for producing concrete blocks *J of Cleaner Production* vol 101 pp 174–179
[15] Drochytka R, et al. 2017 Performance Evaluation and Research of Alternative Thermal Insulation Based on Waste Polyester Fibers *In Procedia Eng* vol. 195 pp 236–243
[16] Trajković D, Jordeva S, Tomovska E and Zafirova K 2017 Polyester apparel cutting waste as insulation material *J of the Textile Institute* vol 108 pp 1238–1245
[17] Gotoh K and Yoshitaka S 2013 Improvement of soil release from polyester fabric with atmospheric pressure plasma jet *Textile Research J* vol 83 pp 1606 –1614

[18] Heniegal A, et al. 2015 Physical and mechanical properties of concrete incorporating industrial and agricultural textile wastes *Int J of Research in Eng and Tech* vol 04 pp 166–176
[19] Deelaman W, et al. 2018 Effect of Banana Fibers on Mechanical and Physical Properties of Light Weight Concrete Blocks *Applied Mechanics and Mat* vol 879 pp 151–155
[20] Paulino J and Espino R 2019 Análisis comparativo de la utilización del concreto simple y el concreto liviano con perlitas de poliestireno como aislante térmico y acústico aplicado a unidades de albañilería en el Perú *Universidad Peruana de Ciencias Aplicadas* pp 1-144
[21] Karpagam V and Jagadeesh K 2018 Study on mechanical properties of concrete blocks by partial replacement of textile waste *Indian J of Env Protection* vol 38 pp 477 481
[22] Patnaik A, et al. 2015 Thermal and sound insulation materials from waste wool and recycled polyester fibers and their biodegradation studies *Energy and Buildings* vol 92 pp 161–169
[23] Valverde I, et al. 2013 Development of new insulation panels based on textile recycled fibers *Waste and Biomass Valorization* vol 4 pp 139–146
[24] American Society for Testing and Materials. ASTM C150. *Standard Specification for Portland Cement.*
[25] American Society for Testing and Materials. ASTM C33M. *Standard Specification for Concrete Aggregates.*
[26] American Society for Testing and Materials. ASTM C192. *Standard Specification for Concrete Aggregates.*
[27] American Society for Testing and Materials. ASTM C685M. *Standard Specification for Concrete Made by Volumetric Batching and Continuous Mixing*
[28] American Society for Testing and Materials. ASTM C90. *Standard Specification for Loadbearing Concrete Masonry Units*
[29] American Society for Testing and Materials. ASTM C173. *Standard Test Method for Air Content of Freshly Mixed Concrete by the Volumetric Method*
[30] American Society for Testing and Materials. ASTM C566. *Standard Test Method for Total Evaporable Moisture Content of Aggregate by Drying*
[31] American Society for Testing and Materials. ASTM C1314-18. *Standard Test Method for Compressive Strength of Masonry Prisms*
[32] Asociación Española de Normalización. UNE-EN ISO 140-5. *Acústica. Medición del aislamiento acústico en los edificios y de los elementos de construcción. Parte 5: Mediciones in situ del aislamiento acústico a ruido aéreo de elementos de fachadas y de fachadas.*
[33] Asociación Española de Normalización. UNE-EN-ISO 10456. *Materiales y productos para la edificación. Procedimientos para la determinación de los valores térmicos declarados y de diseño.*
[34] Arrieta J and Peñaherrera E. 2001 *Fabricación de bloques de concreto con una mesa vibradora.* Centro Peruano Japonés de Investigaciones Sísmicas y Mitigación de Desastres, Lima. pp 1-67
[35] Echevarría E 2017 *Ladrillos de concreto con plástico PET reciclado* Licentiate Thesis Universidad Nacional de Cajamarca, Cajamarca. pp 1-173
[36] Agrawal S, et al. 2013 Utilization of textile apparel waste in clay brick *International Journal of Advanced Research in Engineering and Technology* vol 4 pp 48-52.
[37] Algin H and Turgut P 2008 Cotton and limestone powder wastes as brick material *Construction and Building Materials* vol 22 pp 1074–1080
[38] Rubino C, Liuzzi S, Martellotta F and Stefanizzi P 2018 Textile wastes in building sector: A review *Modelling, Measurement and Control B* vol 87 pp 172-179
[39] Sung C, et al. 2012 Void ratio and durability properties of porous polymer concrete using recycled aggregate with binder contents for permeability pavement *J of Applied Polymer Science* vol 126 pp 338–348
[40] Patidar R and Yadav S 2017 Experimental study of pervious concrete with polypropylene fiber *Int Research J of Eng and Tech* vol 4 pp 22-27.
[41] Carrión F, et al. 2014 Mechanical and Physical Properties of Polyester Polymer Concrete Using Recycled Aggregates from Concrete Sleepers *The Scientific World Journal* vol 2014 *pp* 1–10

[42] Dos Reis J 2009 Effect of textile waste on the mechanical properties of polymer concrete *Mat Research* vol 12 pp 63–67
[43] Mathda V and Khaire H 2016 Study of Effects of Polyester Fibers on Compressive Strength of Concrete *Int J for Research in Applied Science & Eng Tech* vol 4 pp 53-56
[44] Al-Tayeb M, et al. I 2017 Ultimate failure resistance of concrete with partial replacements of sand by waste plastic of vehicles under impact load *Int J of Sustainable Built Env* vol 6 pp 610–616.
[45] Narang P 1995 Material parameter selection in polyester fibre insulation for sound transmission and absorption *Applied Acoustics* vol 45 pp 335–358
[46] Schrader P, et al. 2018 Finite Element Analysis of the Acoustic Behavior of Poroelastic Materials Based on Experimentally Determined Frequency-Dependent Material Properties *Int Conference on Noise and Vibration Engineering* vol 1 pp 15
[47] Watanabe K 1999 Development of high-performance all-polyester sound-absorbing materials. *Society of Automotive Engineers of Japan JSAE Review* vol 20 pp 357–362.
[48] El Wazna M, et al. 2020 Polyurethane coated non-woven: A Promising Solution for Building Insulation *Conference Mat Science and Eng.* vol 827 pp 1-6
[49] Hes L 1999 Optimisation of shirt fabrics' composition from the point of view of their appearance and thermal comfort *Int J of Clothing Science and Tech* vol 11 pp 105–119
[50] Sampath M, et al. 2011 Analysis of thermal comfort characteristics of moisture management finished knitted fabrics made from different yarns *J of Industrial Textiles* vol 42 pp 19–33
[51] Adamczyk J and Dylewski R 2017 The impact of thermal insulation investments on sustainability in the construction sector *Renewable and Sustainable Energy Reviews* vol 80 pp 421-429
[52] Blácido R and Mallqui M 2019 Propuesta de un bloque de concreto con áridos reciclados procedentes del hormigón para la albañilería confinada en Lima Metropolitana Licentiate Thesis *Universidad Peruana de Ciencias Aplicadas* pp 1-126
[53] Izquierdo M and Ortega O 2017 Desarrollo y aplicación del concreto celular a base de aditivo espumante para la elaboración de bloques macizos destinados a tabiquerías no portantes en edificaciones Licentiate Thesis *Universidad Peruana de Ciencias Aplicadas* pp 1-105

IOP Conf. Series: Materials Science and Engineering 1054 (2021) 012006 doi:10.1088/1757-899X/1054/1/012006

Evaluation of the Thermic Efficiency of the Prototype at Scale of a Sustainable Housing that Uses Concrete with PET Fibers (CFP) and the Trombe System

B Dueñas[1], W Soto[1*] and E Carrera[1]

1 Faculty of Engineering, Universidad Peruana de Ciencias Aplicadas, Prolongación Primavera, 2390, Lima, 15023

*u201403860@upc.edu.pe

Abstract. In the last century, climate change has been one of the most notorious consequences of global warming. These changes are manifested through extremely warm or icy temperatures. Being the latter, the one that affects Andean populations in diverse parts of the world. In view of this, many researchers propose various techniques or methods to improve the conditions of habitability and thermic comfort to confront these cold temperatures. This study conducts an experimental investigation of the improvement of thermic efficiency that reaches a prototype of housing to include polyethylene terephthalate (PET), as a component of the concrete mixture in the structural elements. For this purpose, studies were reviewed that allowed selecting the most optimal considerations, with respect to the elaboration and inclusion of PET in the mixture. In addition, the proposed technique includes an adaptation and innovation of the trombe system in the roof, in order to optimize the thermic properties. The results in the present study show that this combination has allowed us to considerably improve the thermal efficiency properties in the prototype of the housing, which is an indicator of its performance when it is applied to real scale housing.

1. Introduction

Global warming and its consequences is an issue of great concern to the world's population. The most notorious consequence is the climate change that occurs in diverse parts of the world through extremely warm or icy temperatures.

These changes have led to various techniques or methods being proposed to mitigate or relatively control extreme temperatures. Romero [1] proposes to implement a photovoltaic system on the roofs of the houses to use the excess of the energy reduction in the generation of cold or heat in the environment. Lizana [2] proposes the use of phase change materials (PCM) that allow the absorption of heat during the day and the release of this during the night. It is also mentioned the improvement of the thermomechanical properties (PTM) by including the glass fiber reinforced polymer (GFRP) in the production of adobe [3] and the plastic in the production of concrete [4].

For this reason, many authors propose plastic, specifically polyethylene terephthalate (PET), as a research material in the production of concrete. The use of this proposed material is given as an addition [5-6], a substitute for cement in the production of lightweight concrete [7] or as a replacement for aggregate [8]. Fraternali [5] concludes that the addition of PET allows for a decrease in the value of thermic conductivity (TC) and an increase in the resistance to compression (RC) in an 18% and 35.14%, respectively, with respect to conventional concrete. Pereira [6] mentions the influence of the

IOP Conf. Series: Materials Science and Engineering 1054 (2021) 012006 doi:10.1088/1757-899X/1054/1/012006

length and volume of PET in the improvement of the resistance to compression (CR) and tensile resistance (TR). Benosman [7] in his experimental research concludes that the inclusion of PET fibers in the cementitious matrix, allows reducing the ultrasonic pulse speed (UPV), thermic conductivity (TC) and absorption capacity (AC) of the compounds. Rodríguez [8] replaces in the concrete mixture to the fine aggregate with PET fibers in percentages of 0%, 4.5%, 5% and 5.5%. The tests performed allowed inferring that the thermal conductivity decreases, thus obtaining a greater degree of thermic insulation.

As a complement, this study carries out an experimental investigation of the influence of PET on the achievement of thermic comfort that can be obtained when combined with an innovative application of the use of the trombe system in the roof. In most of the studies, it is proven that the inclusion of PET allows to conclude indirectly the improvement of thermic insulation (TA). This study will assume the characteristics and volume of PET, which allow the most optimal thermo-mechanical properties (PTM), being these the ones that will be used in the concrete mixture. Likewise, it will be explained in detail how the concept of trombe walls was applied in the elaboration of the roof in a housing prototype. The innovative application of the trombe system includes Dong's proposal [9], which implies the partial use of an aluminium plate to replace the black paint to cover the surface that will capture the solar radiation. This will allow maintaining an adequate temperature for a longer period of time.

2. Methodology

2.1. Materials

The used materials in the present research conform the production of the reinforced concrete, the adaptation of the trombe system in the roof and the construction of the confined masonry system. The first one implies the use of cement (Portland type 1); the used standards for the validation of the used cement is the Peruvian Technical Standard 334.009 [10], which defines the performance requirements of Portland Cement. It also implies the elaboration of PET fibers, in the form of quadratic fractions of approximately 1 cm. The used aggregates in the mixture correspond to HUSO 6 and must comply with the requirements detailed in [11] because they will be exposed to a cold climate. The second one involves the use of a transparent sheet of glass, whose thickness for safety reasons was 8 mm. The black paint will be used on the roof after emptying the slab by capturing the solar radiation and allowing the functionality of the trombe system. In addition, an aluminium plate will be used, which is based on the study of [9]. Technopor cassettes will also be used as a replacement for the roofing brick, due to its contribution to thermic insulation and the ease it will provide us to implement the trombe system on the roof.

2.2. Mixture design

The resistance to be obtained corresponds to a value equal to or greater than 17.5 MPa, due to the fact that the concrete will be used in the confinement elements of the confined masonry system in the one-floor housing prototype. In addition, it is desired to obtain a 100 mm SLUMP to facilitate the workability of the concrete. The used method to perform the mixture design is the ACI method.

The inclusion of PET fibers in the mixture, as a replacement for fine aggregate, corresponds to a percentage of 4%. This percentage designated for the inclusion of PET is based on studies previously carried out; in which Rodríguez [8], when presenting a mixture design similar to ours, will guarantee the fulfillment of the minimum resistance considered in the design and an improvement in thermic insulation. The mixture design is shown in table 1, which shows the weight of each material for a cubic meter of concrete.

Table 1. Mixture design with PET fibers.

Components of the mixture	Unit	Added Sample
Portland Cement I	kg/m^3	367
PET fibers	%	4
	kg/m^3	67.4
Water	lt	177.3
a/c ratio	-	0.50
Sand	kg/m^3	723.7
Stone	kg/m^3	834.1

The figure 1 shows the concrete mix reinforced with PET fibers, which will be used in the columns, beams and joists of the prototype.

Figure 1. Concrete mix with PET fibers.

2.3. *Prototype design*

A design was made in the AutoCAD program to have a visualization in plant of the details to be considered in the prototype. The figure 2. presents the dimensions of the prototype and the location of the door and window.

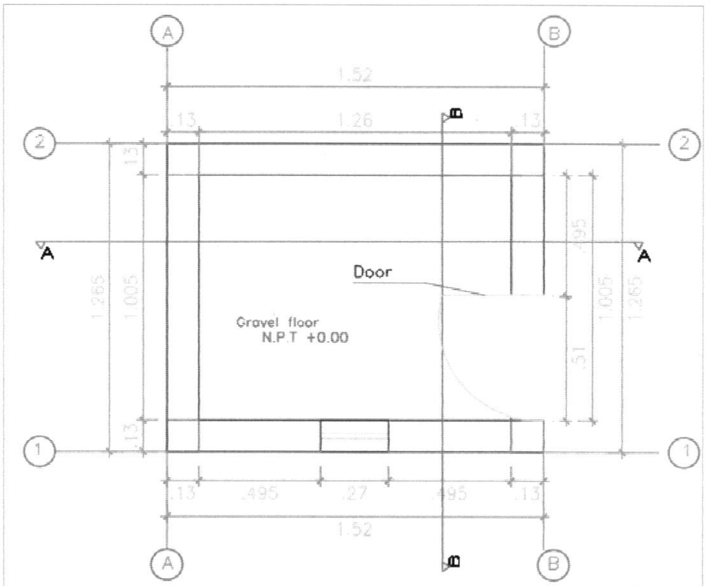

Figure 2. Detail in plant of the housing prototype.

Figure 3 and Figure 4 show the considerations of the construction process of the lightened slab and implementation of the trombe system on the roof. The first one will consider making holes in the technopor cassettes and ensuring their continuity throughout the thickness of the roof. The second one will consider the painting of the surface after casting and the way the aluminium sheet will be used.

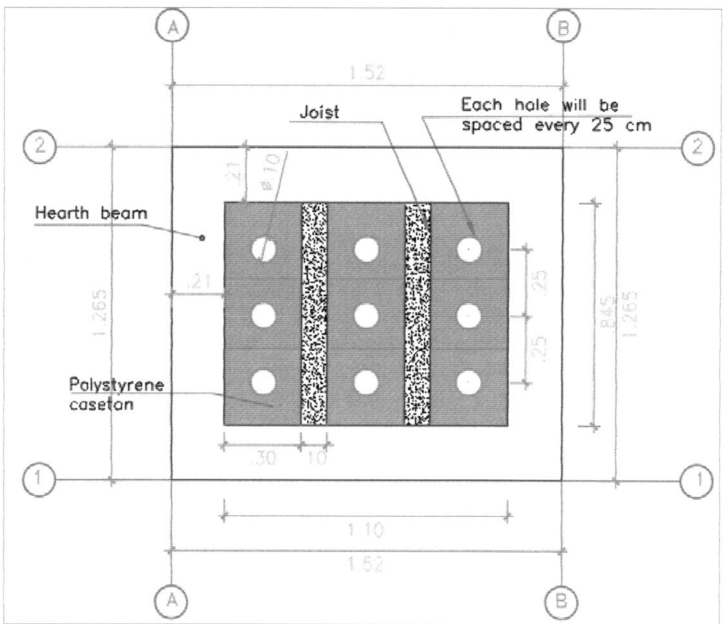

Figure 3. View in plant of pre-casting lightened roof.

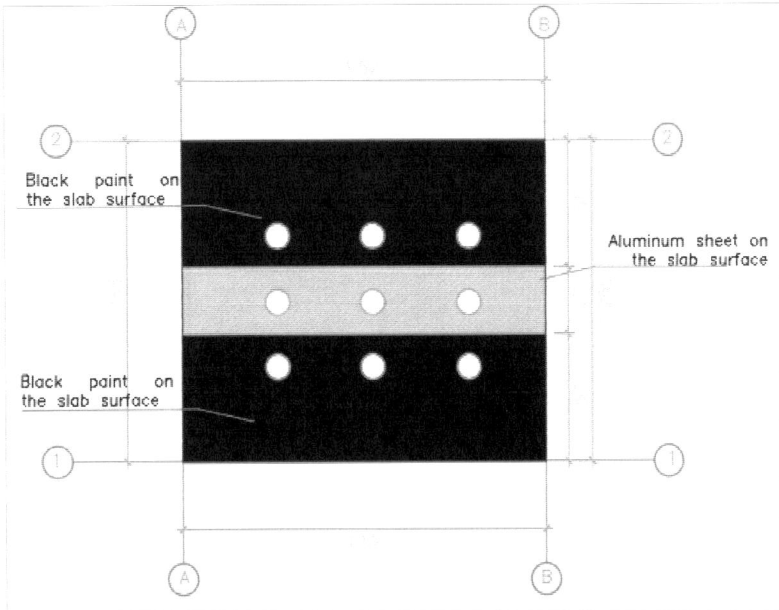

Figure 4. View in plant of post-casting trombe roof.

2.4. Constructive process

Once the technical specifications that the housing prototype was going to present were detailed, we proceeded to build it, obtaining the prototype as shown in figure 5. Also, figure 6 allows visualizing the proposal to implement the trombe system on the roof.

Figure 5. Final physical prototype of the proposal.

Figure 6. Adaptation of the trombe system on the roof.

3. Results and discussion

3.1. Temperature

Figure 7 and Figure 8 show the temperature trend of a typical sunny and cloudy day, respectively, during the recording period. This will allow to visualize the differences that occur between the ambient temperature (AT) and the temperature inside the prototype of the housing (TP).

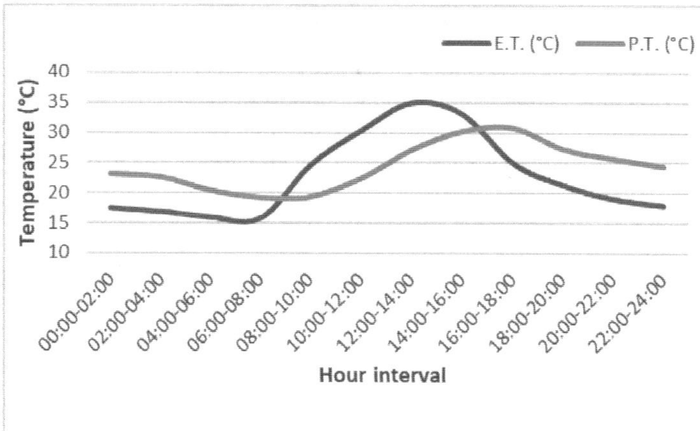

Figure 7. Temperature graph of a typical sunny day.

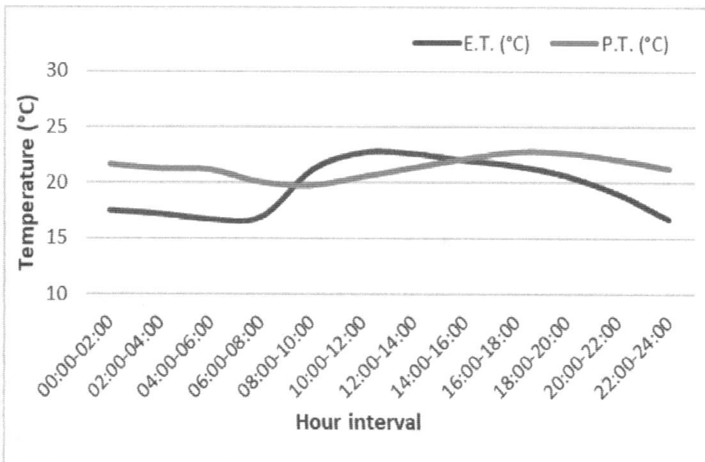

Figure 8. Temperature graph of a typical cloudy day.

On a cloudy day the greatest difference in temperature between the environment and the interior of the house was 4.6 °C, which occurred in the interval from 10 to 12 midnight. This means that the prototype has a temperature improvement of 27.5 % with respect to the environment. On a sunny day, the greatest difference in temperature between the environment and the interior of the housing was 6.5 °C, which occurred in the interval from 10 to 12 midnight. This means that the prototype has a temperature improvement of 36.1 % with respect to the environment. Likewise, it is inferred that this percentage will increase as the day presents more solar radiation.

3.2. Relative humidity

Figure 9 and Figure 10 show the trend in relative humidity on a typical sunny and cloudy day, respectively, during the recording period. This will allow to visualize the differences between the ambient humidity (HA) and the humidity inside the prototype of the housing (HP).

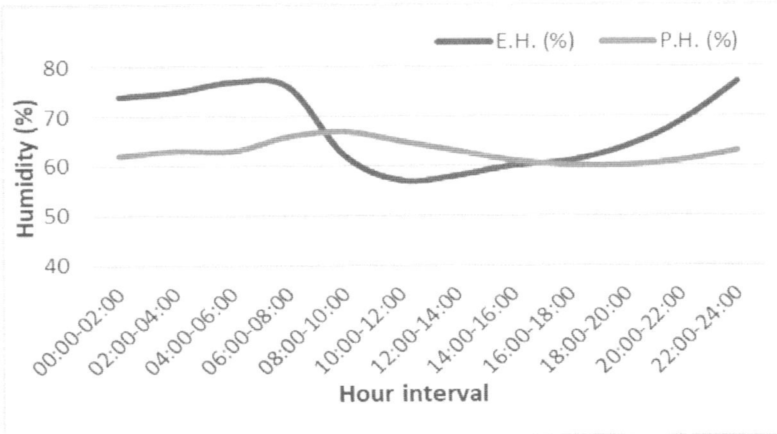

Figure 9. Relative humidity graph of a typical sunny day.

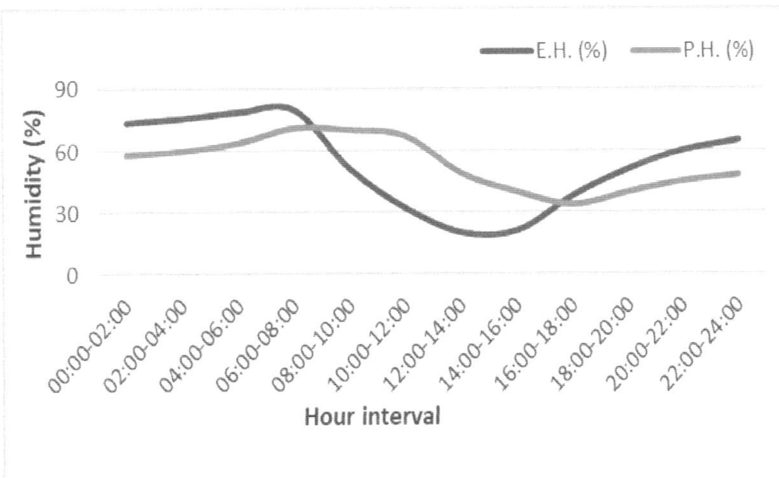

Figure 10. Relative humidity graph of a typical cloudy day.

The trend of the graphs that represent the relative humidities allows to infer that the variation of these parameters is inversely proportional to the variation of the temperatures.

3.3. CO_2 Reduction

Figure 11 contrasts the difference, in terms of CO_2 production, between reinforced concrete with PET fibers and the conventional concrete. This will highlight the importance of this type of reinforced concrete as a contribution to environmental impact.

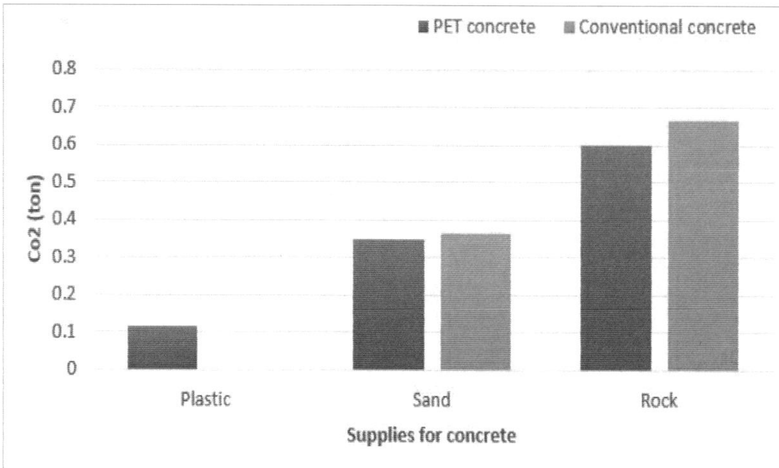

Figure 11. CO_2 emission graph between PET concrete vs. conventional concrete.

The graph shows that using concrete with PET fibers benefits the environmental impact in a positive way, because with conventional concrete more sand and stone input is used and therefore the CO_2 emission increases. The relevance of the decrease in CO_2 will be considerable as the magnitude of the work increases. Although the replacement of aggregates contributes to this decrease, it is important to point out the new use that will be given to plastic bottles, which are currently the materials that produce the most environmental pollution.

4. Conclusions

Based on the obtained results in the experimental part of this study, the following conclusions are shown:

- It was demonstrated that the combination of the trombe system in the roof with the use of a reinforced concrete with PET fibers allowed to improve the thermic insulation in the prototype of the housing.
- On the cloudy day, the greatest difference in temperatures was 4.6 ° C (27.5%), which occurred at midnight. It is inferred that the tempered air during the day is partially released during the night.
- On the sunny day, the largest temperature difference was 6.5 ° C (36.1%), which occurred at midnight. It is inferred that this percentage will increase as the day presents more solar radiation.
- On average during the 8 a.m. span at 2 p.m., the temperature in the home increased as the ambient temperature reached its highest peak. After 2 p.m., the ambient temperature begins to drop rapidly. However, the temperature of the house still increased relatively until 4 p.m., from this time the temperature begins a decrease, not as abrupt as that of the environment. This difference in temperatures is conserved during the night and early the next day until 8 a.m., at which time the described cycle begins again.
- Although the aim was to demonstrate that the applied technique would allow the isolation of cold temperatures, it was demonstrated that it was also possible to isolate the high temperatures of the study area, generating percentage differences of up to 25.4%.
- The implementation of the aluminium plate, in conjunction with the black painted surfaces, made it possible to optimize solar capture and induce an improvement in the functioning of the trombe system on the roof.
- The use of reinforced concrete with PET fibers had an incidence in terms of environmental impact, because when carrying out an analysis, it was concluded that the use of this type of concrete reduced CO_2 emissions.

IOP Conf. Series: Materials Science and Engineering 1054 (2021) 012006 doi:10.1088/1757-899X/1054/1/012006

5. References

[1] Romero Rodríguez, L., Sánchez Ramos, J., Guerrero Delgado, M., Molina Félix, J., & Álvarez Domínguez, S. (2018). Mitigating energy poverty: Potential contributions of combining PV and building thermal mass storage in low-income households. *Energy Conversion and Management, 173*, 65-80.

[2] Lizana, J., de-Borja-Torrejon, M., Barrios-Padura, A., Auer, T., & Chacartegui, R. (11 de 2019). Passive cooling through phase change materials in buildings. A critical study of implementation alternatives. *Applied Energy, 254*, 113658.

[3] Gandia, R., Gomes, F., Corrêa, A., Rodrigues, M., & Mendes, R. (20 de 10 de 2019). Physical, mechanical and thermal behavior of adobe stabilized with glass fiber reinforced polymer waste. *Construction and Building Materials, 222*, 168-182.

[4] Záleská, M., Pavlíková, M., Pokorný, J., Jankovský, O., Pavlík, Z., & Černý, R. (2018). Structural, mechanical and hygrothermal properties of lightweight concrete based on the application of waste plastics. *Construction and Building Materials, 180*, 1-11.

[5] Fraternali, F., Ciancia, V., Chechile, R., Rizzano, G., Feo, L., & Incarnato, L. (2011). Experimental study of the thermo-mechanical properties of recycled PET fiber-reinforced concrete. *Composite Structures, 93*(9), 2368-2374.

[6] Pereira, E., de Oliveira Junior, A., & Fineza, A. (2017). Optimization of mechanical properties in concrete reinforced with fibers from solid urban waste (PET bottles) for the production of ecological concrete. *Construction and Building Materials, 149*, 837-848.

[7] Benosman, A., Mouli, M., Taibi, H., Belbachir, M., Senhadji, Y., Bahlouli, I., & Houivet, D. (2017). *CHEMICAL, MECHANICAL AND THERMAL PROPERTIES OF MORTAR COMPOSITES CONTAINING WASTE PET.*

[8] Rodrigues, N., Carvalho, M., Balbino, A., Vasconcelos,A. (2018). *PET FIBER IN CONCRETE PRODUCTION.*

[9] Dong, J., Chen, Z., Zhang, L., Cheng, Y., Sun, S., & Jie, J. (1 de 2 de 2019). Experimental investigation on the heating performance of a novel designed trombe wall. *Energy*, 728-736.

[10] NTP 334.009:2016, "CEMENTO. Cemento Portland requisitos", INACAL, Lima, 2016.

[11] NTP 400.012:2013, "AGREGADOS. Análisis granulométrico del agregado fino, grueso y global", INACAL, Lima, 2013.

A Self-defined Index for Hygrothermal Performance-Oriented Development of Bamboo and Its Test Method

Zujian Huang[1,2,*], **Yimin Sun**[1,2,3]

1. School of Architecture, South China University of Technology, Guangzhou 510640, China;
2. State Key Laboratory of Subtropical Building Science, South China University of Technology, Guangzhou 510640, China;
3.Guangdong Research Center of Sustainable Architecture and Urban Design Enginee ring Technology, Guangzhou 510640, China

*huangzuj@scut.edu.cn

Abstract. In the manufacturing process of the mechanical performance-oriented bamboo products, the bulk density is often used as an important control index since it has significant correlation with the mechanical properties. However, the development of bamboo for building envelope is supposed to be hygrothermal performance-oriented, which may require a different control index. In this study, 6 typical bamboo products are selected as the research objects. Based on the material parameter requirements of the building envelope coupled heat and moisture process model, 6 test items for hygrothermal properties and 2 test items for basic properties, including the dry bulk density (ρ_{d}) and a self-defined 'open porosity of liquid water (φ_{l})', are carried out. Correlation analyses between the tested hygrothermal properties and the basic properties show that, macroscopically both moisture and heat transport properties show stronger correlation with φ_{l} than that with ρ_{d}. Taking the correlation coefficient (r) and the statistical significance (*sig*) as indicators, the correlations between the hygrothermal properties and φ_{l} are, respectively, 107.23%–140.00% and 1.44%–11.94% of that between the hygrothermal properties and ρ_{d}. The phenomenon proves that φ_{l} is potential to be a better control index for the development of hygrothermal performance-oriented bamboo. Based on this, a method including 6 operation steps to test the φ_{l} is introduced.

1. Introduction

Bamboo is a fast-growing plant, distributed in the low latitudes in Asia Pacific, America and Africa, and is widely used in the local traditional constructions. Nowadays, the promotion of 'building with bamboo' has multiple potentials such as the high-value utilization of the bamboo resource and sustainable development of the building industry. Under the background of industrialization of the bamboo forest resources, a variety of bamboo products have been developed, including the plybamboo (bamboo plywood), bamboo particle board, bamboo oriented strand board, bamboo laminated lumber, bamboo scrimber, etc. [1] which are used in the load-bearing structural elements such as beams and columns, as well as in the non-load-bearing building envelope elements such as the walls and floors, etc.

The mechanical strength of certain materials largely depends on the solid part, which results in a significant correlation between the mechanical properties and the bulk density. Therefore, for a long period, bulk density is commonly used as a manufacturing indicator for the control and strength prediction of those mechanical performance-oriented building materials, including the wood/bamboo-based panels [2]. Other than this, those materials used in building envelope is oriented by hygrothermal performance. As a fundamental work, the material tests on bamboo for different property parameters have been widely carried out.

1.1. Overview of the bamboo properties studies

The material parameters have important impacts on the performance of building envelope, such as the heat and moisture flows through the exterior wall, the indoor environment and the building energy consumption [4]. Previous studies on hygrothermal properties of bamboo mostly focus on raw bamboo, measuring the specific heat capacity, thermal conductivity, sorption isotherm and water vapor transmission resistance factors, etc. [4]

The Moso bamboo (*Phyllostachys pubescens*), a common bamboo species, is often used as the test object in these studies. The tests on the specific heat capacity of Moso bamboo result in various values from 1.7 to 2.3 kJ/(kg·K) [4] at 40°C by the differential scanning calorimetry method. The raw bamboo is proved to be significantly non-homogenous in terms of the thermal properties in the longitudinal, radial and tangential directions. The thermal diffusivity of Moso bamboo in longitudinal direction, as well as the thermal conductivity of Moso laminated board and bamboo veneer board have also been measured [5].

The sorption isotherm of bamboo blocks, bamboo powder, parenchyma cells and chemically macerated fibers are tested by different scholars, and the comparison results show similar laws that the hygroscopic capability are, in descending order, the parenchyma cells, the bamboo fibers, the bamboo powder and the bamboo blocks [6]. Different values have been measured from the samples taken from different parts of a bamboo wall. For the liquid water-related properties, the capillary saturated moisture contents are 572, 479 and 385 kg/m^3, respectively, for the exterior, middle and interior parts of the bamboo culm wall, and the corresponding liquid water absorption coefficients are 0.014, 0.008 and 0.0019 kg/(m^2·s$^{0.5}$) [4]. The dry cup test on the water vapor diffusion resistance factor shows the values in the radial and tangential directions range from 30 to 57 [4].

Bamboo wall, as the main part for industrial utilization, has been studied by scholars from disciplines of botany, material science and bamboo-based panels production, etc. for the microstructure, surface morphology, density distribution, pore structure, cracking characteristics, etc. [7,8] The density distribution observation by computer tomography scanning shows that the bulk density of Moso bamboo nodes and internodes is in the range 600-800 kg/m^3 [4]. When taking the bamboo-based panels into account, the bulk density of bamboo products is from 650 to 1290 kg/m^3, which is substantially higher than timber products that are generally in the range 400-800 kg/m^3. An observation on the pore structure of bamboo culms by backscattered electron scanning and computerized tomography shows that the average porosity of Moso bamboo internode is between 44.9% and 63.4% [4].

1.2. Hypothesis and motivation of this study

In the heat and moisture process model of building envelope, a variety of parameters that characterize the materials' properties are involved, which can generally be classified as the basic properties, hygric properties and thermal properties. The hygric and thermal properties can be further divided into parameters characterizing the storage and transfer properties. The tests on hygrothermal properties is extremely slow, which take months or even years to obtain the complete results. Therefore, finding a basic physical property indicator that is easy to obtain and can accurately predict those hygrothermal properties is of importance [3].

Basic properties can be categorized into two groups: the skeletal features that characterize the solid sections and the pore features that characterize the internal space. For most physical properties related to the heat and moisture process, except the thermal storage property (specific heat capacity) that is depending on the chemical composition, the storage locations and transport channels of the moisture (including liquid and gaseous water) as well as a part of the heat transport occurs in the internal space. Therefore, it can be considered that using bulk density to predict the hygric and thermal properties would have inevitable limitations.

As a plant fiber-based organic material, bamboo has a complex pore structure, so it is not feasible to be described on a microscopic level. 'Porosity', as an index to describe the ratio of the internal space to the whole volume, is relatively easy to obtain. However, the pores within bamboo are in two statues, namely the open pores and closed pores, the latter of which is isolated from the outside, so that moisture absorption and desorption processes hardly occur there. However, in general testing process, the obtained 'porosity' includes both the open pores and closed pores.

'Open porosity' is used to describe the part of space where moisture storage and transport can occur. However, the results measured by general vacuum saturation experiments include both liquid water and gaseous water in total amount, which cannot be distinguished. Although the latter has a smaller mass proportion, there is a difference in the initial moisture content, ie, the gaseous water content, between various bamboo samples, which is normally in the range 8%-15% [9,10]. This causes errors to the test results and prevents the test method from forming a uniform standard.

The aim of this study is to propose a method to test the volume of the open pores and exclude the interference of the initial gaseous water content, and verify its possibility to be a new index for the hygrothermal performance-oriented development of bamboo.

2. Material and method

2.1. Test object

Based on the investigation of literature, material manufacturers and the classification of representative bamboo products, 6 kinds of test samples are selected, including the flattened bamboo panel (FB), plybamboo (BMB), bamboo particleboard (BPB), bamboo oriented strand board (BOSB), bamboo laminated lumber (BSB) and bamboo scrimber (BFB). Among these samples, FB represents the raw bamboo that is flattened without applying glue, and the rest are bamboo-based panels. (figure1)

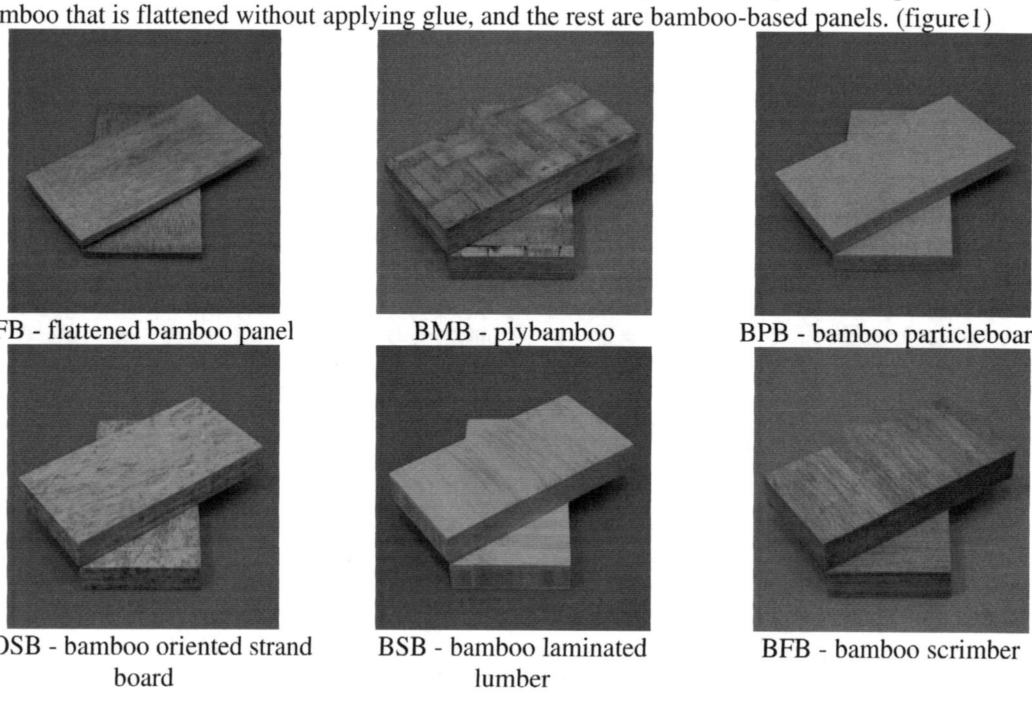

Figure 1. Photos of the material samples.

2.2 Test method

2.2.1 Hygrothermal properties. To obtain the necessary hygrothermal properties, six test items are carried out, including the sorption test for the moisture storage properties; capillary absorption test, water vapor transmission test and drying test for the moisture transport properties; thermal analysis for the heat storage properties, and thermal conductivity test for the heat transport properties. (table 1)

Table 1. Hygrothermal properties test items and the operation methods.

Category	Test items	Target value	Reference of the operation methods
Hygric properties	1. Sorption test	20°C isothermal adsorption and desorption curve	The ISO standard: ISO 12571:2012, and the American standard: ASTM-04a(2016) [11,12]
	2. Capillary absorption test	Water absorption coefficient; Capillary saturated moisture content	The ISO standard: ISO 15148:2002(E) [13]
	3. Water vapor transmission test	Water vapor transfer coefficient	The ISO standard: ISO 12572:2001(E), and the American standard: ASTM E 96/E 96M-2005 [14,15], in which the water vapor transfer coefficient of still air δ_{air} [kg/(m·s·Pa)] at a certain temperature T and pressure p refers from [16]
	4. Drying test	Drying curve; Drying rate curve	The test method from Fraunhofer IBP (accredited by the DIN EN ISO/IEC 17025)
Thermal properties	5. Thermal analysis	Specific heat capacity	The ISO standard: ISO 11357-4-2005 [17]
	6. Thermal conductivity test	Thermal conductivity	The ISO standard: ISO 8302-1991 [18]

2.2.2 Open porosity of liquid water. The study proposes a method for testing the self-defined index, called the 'open porosity of liquid water', for bamboo, which mainly includes the following steps: specimen preparing, specimen drying, dry specimen weighing, specimen vacuum saturating (air pumping, water injecting, and water immersing), vacuum saturated specimen weighing, and the data processing. (figure 2)

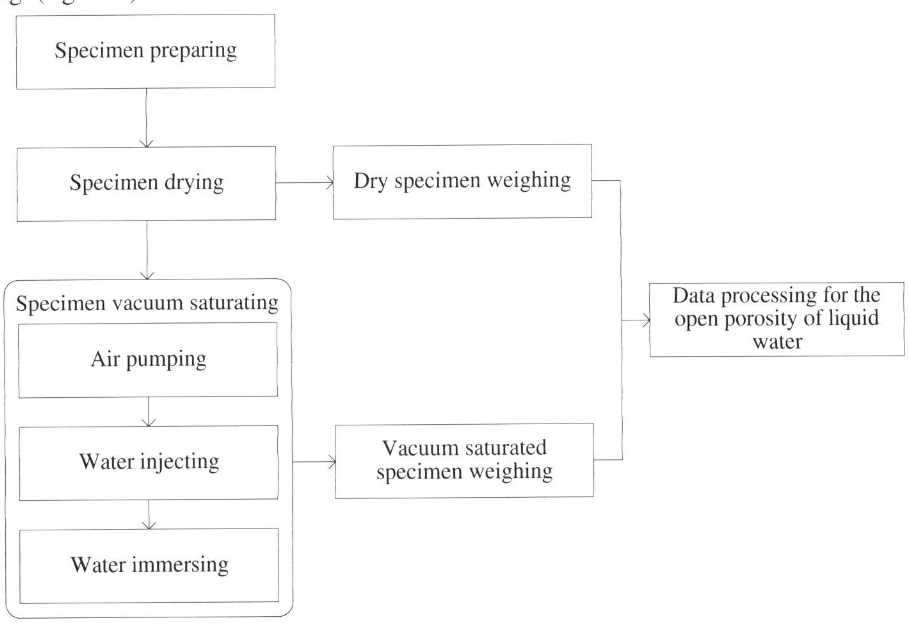

Figure 2. The test flow chart of the 'open porosity of liquid water.

3. Result analysis

3.1 Test results

The test results include the dry bulk density (ρ_d) and open porosity of liquid water (Φ_l) that characterize the basic properties of the test samples; the 20°C isothermal adsorption and desorption curve and the capillary saturated moisture content, water absorption coefficient, water vapor transfer coefficient and the drying rate curve that characterize the hygric properties; the specific heat capacity, 24h thermal storage coefficient and the thermal conductivity that characterize the thermal properties. (table 2, 3)

Table 2. Test results of the basic properties.

Test items	Target value	Abbreviation	Test and calculation results					
			FB	BMB	BPB	BOSB	BSB	BFB
1. Bulk density test	Dry bulk density	ρ_d [kg/m^3]	666.38	776.21	623.32	895.02	563.81	1108.77
2. Open porosity test	Open porosity of liquid water	Φ_l [-]	52.24	49.58	63.17	42.32	53.97	17.36

Table 3. Test results of the hygrothermal properties.

Test items	Target value	Abbreviation	Test and calculation results					
			FB	BMB	BPB	BOSB	BSB	BFB
1. Sorption test	20°C isothermal sorption curve	$w_{RH=50\%}$ [kg/m^3]	31.07	39.20	36.89	46.07	30.44	28.17
2 Capillary absorption test	Water absorption coefficient	A_{cap} [E^{-04} kg/(m^2·s$^{0.5}$)]	74.06	38.71	447.78	59.15	78.74	8.73
	Capillary saturated moisture content	w_{cap} [kg/m^3]	326.41	221.44	521.95	342.68	317.21	165.93
3. Water vapor transmission test	Water vapor transfer coefficient	$\delta_{RH=50\%}$ [E^{-13} kg/(m·s·Pa)]	27.81	19.28	93.28	7.23	27.99	4.37
4. Drying test	Drying rate (*T*=23°C, *RH*=50%)	$U_{u12\%\text{-}u6\%}$ [E^{-07} kg/(m^2·s)]	135.15	18.32	126.46	38.01	60.20	10.14
5. Thermal analysis	Specific heat capacity	c [J/(kg·K)]	1796	2020	1760	1663	1960	1550
	24h thermal storage coefficient	S_{24h} [W/(m^2·K)]	5.93	8.57	4.87	6.94	6.63	8.68
6. Thermal conductivity test	Thermal conductivity	λ [W/(m·K)]	0.1088	0.1733	0.0801	0.1197	0.1475	0.1625

3.2. Correlation anaylsis

Correlation analyses of the test results are carried out between the basic properties (ρ_d, Φ_l) and the hygrothermal properties. Results show that the correlation between the Φ_l and the hygrothermal properties is stronger than that between the ρ_d and the hygrothermal properties, which proves that the Φ_l would be a better indicator to predict the hygrothermal properties of bamboo. The comparison results could be expressed as the ratio of the correlation coefficient (r) and the statistical significance (*sig*) of the two groups. The correlation strength is proportional to the *r* value and inversely proportional to the *sig* value.

The correlation between the hygrothermal properties with ϕ_l is, respectively, 107%-140% (*r*) and 1.4%-11.9% (*sig*), of the correlation between the hygrothermal properties and ρ_d. Compared with the dry bulk density, the correlation strength of the open porosity of liquid water has been significantly improved. (table 4)

Table 4 Correlation anaylsis results between the basic properties and hygrothermal properties.

Items	Correlation index	w_{cap}	δ	A_{cap}	U	c	λ
ρ_d	correlation coefficient (*r*)	0.6740	0.6745	0.9002	0.7891	0.4857	-0.5925
	statistical significance (*sig*)	0.002159	$1.4681E^{-7}$	$3.5973E^{-7}$	$1.0776E^{-8}$	0.328723	0.009565
ϕ_l	correlation coefficient (*r*)	-0.5361	-0.6014	-0.8249	-0.7359	-0.6000	0.4232
	statistical significance (*sig*)	0.021829	0.000006	0.000025	$3.1545E^{-7}$	0.208000	0.080111
r-ϕ_l/ρ_d		125.72%	112.15%	109.13%	107.23%	-	140.00%
sig-ϕ_l/ρ_d		9.89%	2.45%	1.44%	3.42%	-	11.94%
number of samples (N)		18	48	18	36	-	18

Note: since the *c* depends on the chemical composition of the material, it is excluded from the correlation analysis.

4. Conclusion

1) The 'open porosity of liquid water (ϕ_l)' proposed in this study is potential to be a basic property control index for the hygrothermal performance-oriented development of bamboo.

As a simplified indicator, ϕ_l is easy to measure, and costs much shorter time than the testing of hygrothermal properties. For the application of bamboo in building envelope, due to the significant correlation between the ϕ_l and the hygrothermal properties, it can provide an efficient and accurate prediction for the selection of material products in practical engineering. For the production of bamboo products, ϕ_l is potential to be a hygrothermal performance-oriented production control index.

2) The test procedure for ϕ_l.

A method proposed in this study for testing the ϕ_l of bamboo includes the following steps: 1) Specimen preparing: make a plurality of specimens (suggested to be 6 pieces) of standard size; 2) Specimen drying: dry the specimens from the step 1 to obtain dry specimen; 3) Dry specimen weighing: weigh the specimens and record the weight of the dry specimens; 4) Specimen vacuum saturating: place the dry specimens into the sink in the vacuum chamber, close the door of the chamber, pump out the air from the chamber and the specimen, then start inject water into the sink and keep the specimens immersed in water to obtain the vacuum saturated specimens; 5) Vacuum saturated specimen weighing: weigh the specimens and record the weight of the vacuum saturated specimens; 6) Data processing.

5. References

[1] Huang Z, Sun Y and Musso F 2017 Assessment of bamboo application in building envelope by comparison with reference timber *Construction and Building Materials* 156 pp 844-860

[2] Sumardi I, Kazuhiro O and Shigehiko S 2007 Effect of board density and layer structure on the mechanical properties of bamboo oriented strandboard *Journal of wood science* 53(6) pp 510-515

[3] Huang Z 2019 *Application of Bamboo in Building Envelope* (Cham, Switzerland: Springer International Publishing) chapter 2 pp 81-118

[4] Huang P 2017 *Hygrothermal performance of Moso bamboo-based building material* (Diss. University of Bath)

[5] Shah D U, et al. 2016 Thermal conductivity of engineered bamboo composites *Journal of Materials Science* 51(6) pp 2991-3002
[6] Jiang Z, et al. 2012 Hygroscopic behavior of bamboo and its blocking units *Journal of Nanjing Forestry University (Natural Sciences Edition)* 36(2) pp 11-14
[7] Kim J S, et al. 2008 Micromorphological characteristics and lignin distribution in bamboo (Phyllostachys pubescens) degraded by the white rot fungus Lentinus edodes *Holzforschung* 62(4) pp 481-487
[8] Malanit P, et al. 2008 Macroscopic aspects and physical properties of Dendrocalamus asper Backer. for composite panels *Journal of Bamboo and Rattan* 7(3&4) pp 151-163
[9] Huang Z, Sun Y and Musso F 2017 Experimental study on bamboo hygrothermal properties and the impact of bamboo-based panel process Construction and Building Materials 155 pp 1112-1125
[10] Huang Z, Sun Y and Musso F 2019 Hygrothermal performance optimization on bamboo building envelope in Hot-Humid climate region Construction and Building Materials 202 pp 223-245
[11] ISO 12571:2012: *Hygrothermal performance of building materials and products - Determination of hygroscopic sorption properties* (Geneva: International Organization for Standardization)
[12] ASTM-04a(2016): *Standard Test Method for Hygroscopic Sorption Isotherms of Building Materials* (West Conshohocken: ASTM International)
[13] ISO 15148:2002(E): *Hygrothermal performance of building materials and products - Determination of water absorption coefficient by partial immersion* (Geneva: International Organization for Standardization)
[14] ISO 12572:2001(E): *Hygrothermal performance of building materials and products - Determination of water vapour transmission properties* (Geneva: International Organization for Standardization)
[15] ASTM E 96/E 96M-2005: *Standard Test Methods for Water Vapor Transmission of Materials* (West Conshohocken: ASTM International)
[16] Mukhopadhyaya P, Kumaran M K and Lackey J 2005 Use of the modified cup method to determine temperature dependency of water vapor transmission properties of building materials *Journal of Testing and Evaluation* 33(5) pp 316-322
[17] ISO 11357-4-2005: *Plastics - Differential scanning calorimetry (DSC) - Part 4: Determination of specific heat capacity* (Geneva: International Organization for Standardization)
[18] ISO 8302-1991: *Thermal insulation; determination of steady-state thermal resistance and related properties; guarded hot plate apparatus* (Geneva: International Organization for Standardization)

Acknowledgments

The research project is funded by: the National Natural Science Foundation of China (51908219), the China Postdoctoral Science Foundation (2020T130209, 2018M640782), the Natural Science Foundation of Guangdong Province, China (2019A1515012124), the China Association for Science and Technology (B4200030), and the Fundamental Research Funds for the Central Universities (D2181640).

IOP Conf. Series: Materials Science and Engineering 1054 (2021) 012008 doi:10.1088/1757-899X/1054/1/012008

Concrete Cracking Control in *Underwater Marine Structures* using Basalt Fiber

C Quispe[1*], D Lino[1], J Rodríguez[2] and A Hinostroza[3]

[1] Bachelor, Civil Engineering Program, Universidad Peruana de Ciencias Aplicadas, Lima, Perú
[2] Full Professor, Civil Engineering Program, Universidad Peruana de Ciencias Aplicadas, Lima, Perú
[3] Professor, Civil Engineering Program, Universidad Peruana de Ciencias Aplicadas, Lima, Perú

*u201310600@upc.edu.pe

Abstract. The construction of coastal ports requires the use of materials that meet the demands of the marine environment, to prevent underwater concrete structures from cracking and spalling easily; basalt fiber is used to delay the expansion of concrete and prevent the formation of cracks. This research studies the behavior of concrete for prefabricated piles with Portland Cement Type I and basalt fibers added in 0.1%, 0.3% and 0.6%; the results indicate that the fiber is suitable for concrete, the slump decreases, the compressive strength increases for specimens cured in tap water and sea water, the relationship between resistances does not vary, and the depth of carbonation decreases.

1. Introduction

Constructions located in marine environments are exposed to attack by different agents, such as: sulfates, acids, seawater, chlorides, microorganisms, algae and mollusks, so they require the use of a concrete with greater durability [1], which presents low permeability to reduce the appearance of cracks (micro and macro) and voids produced during its manufacturing and commissioning [2]. This reduction has been carried out through the application of fibrous materials such as: glass fibers, polypropylene, steel, and carbon, among others; because with them compact and more efficient concretes are produced [3], being these suitable for marine infrastructures, as is the case of tourist ports located on the beaches of coastal coastlines.

Ports are built to generate economic development in society, but due to social needs for recreation or tourism they are remodeled, including stalls and modern ornaments [4], which increase existing loads, requiring that their supporting structural elements, such as the reinforced concrete piles, are in a good state of preservation to guarantee the strength of the structure. These piles, being submerged under water, are affected by microorganisms, salts and chlorides, the latter producing an increase in cracks in the covering [5]; the same ones that cause cracks and spallings, leaving the reinforcing steel exposed [6] and affecting the load capacity due to the reduction of its cross section [7].

An alternative solution to the reduction of cracking consists of reinforcing the concrete with Basalt Fiber (BF), which comes from an igneous rock and that through a winding process fibers of Ø 5-20 μm; its characteristics include: rigidity, durability, high resistance to acids and solvents, contributes to workability, increases tensile strength, is resistant to chloride attack and is more tenacious in alkaline environments [8]; and that when incorporated into the mix it requires avoiding the formation of agglomerations so as not to affect the final strength of the concrete [8]. Interesting investigations have been reported on the characteristics of BF and its behavior in concrete, thus we have [9], [10], [11] analyzed the chemical composition of BF by means of the X-ray fluorescence test and found that 53% corresponds to silicon dioxide; on the other hand, with the addition of BF the slump decreases [12], [13], [14]; the compressive strength for concrete cured in tap water and sea water increases [14], [15], [16], [17], [18], [19]; the relationship of the compressive strength of concrete cured in sea water and concrete in tap water is similar [20], [21], [22]; and the depth of carbonation decreases [23], [24], [25].

The present work is related to the tourist ports in the bay of the city of Pisco where there are reinforced concrete piles with different pathologies on their surface, highlighting cracking and spalling; Portland Cement Type I has been used because the type of cement is of little importance compared to the low permeability requirements of concrete in water [26]. This research addresses the problem of surface cracking of precast reinforced concrete piles of tourist ports submerged in tidal zones; to do this, we study the chemical composition of the BF, the slump, the compressive strength of the cured concrete in tap water and sea water, the relationship between the cured with sea water and cured with tap water, and the depth of carbonation.

2. Materials and methods

2.1 Materials

It was used Portland cement type I [27]; the fine and coarse aggregates were natural, the coarse ones being a maximum size of Ø ½" [28]; the BF used are circular in shape, of Ø 16 µm, length 24 mm and density 2650 Kg/m^3, the image of which is shown in figure 1; and the water was tap [29].

Figure 1. Basalt fiber.

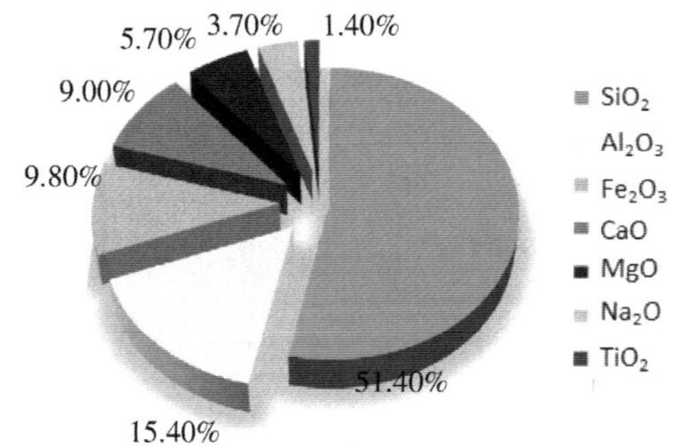

Figure 2. Chemical composition of BF.

2.2 Methods

The concrete design was made according [30] to a f'$_c$= 280 kg/cm^2, The different mixtures made with their component materials, tests and number of test tubes used are shown in table 1 and table 2. The mixture was prepared according to [31], then they were cured for 7 and 28 days [32] and then they were tested in a fresh and hardened state. The composition of the BF was made at 3 days with the X-ray fluorescence assay [33]; the slump with [34]; for resistance specimens of Ø 15 x 30cm [35]; and the depth of carbonation with phenolphthalein in specimens of Ø 15 x 30cm curing with sea water [36].

Table 1. Mix design for 1m^3 of concrete.

Materials (kg)	**BF-0** standard	**BF-0.10** 0.1% BF	**BF-0.30** 0.3% BF	**BF-0.60** 0.6% BF
Type I cement	415.05	415.05	415.05	415.05
Fine Aggregate	764.59	764.59	764.59	764.59
Coarse Aggregate	1010.55	1010.55	1010.55	1010.55

Basalt Fiber	-	0.16	0.50	1.00
Water	194.15	194.15	194.15	194.15

Table 2. Number of specimens per test.

Test type	BF-0		BF-0.10		BF-0.30		BF-0.60		Type of curing
	Curing time (days)	Number of specimens	Curing time (days)	Number of specimens	Curing time (days)	Number of specimens	Curing time (days)	Number of specimens	
Compressive strength	7	4	7	4	7	4	7	4	Tap water, Sea water
	28	4	28	4	28	4	28	4	
Carbonation depth	90	-	90	2	90	2	90	2	Sea water

3. Results and analysis

3.1 X-ray fluorescence
In figure 2, the chemical compounds of the BF of 24 mm length and Ø 16 µm are presented, showing that Silicon dioxide (SiO_2), Aluminum oxide (Al_2O_3), Iron oxide (Fe_2O_3) and Oxide of calcium (CaO) are the 4 compounds that represent 85.60% of the fiber composition, where the amount of SiO_2 is equal to 51.40%. [37] and [38] study BF of 21mm and 25mm in length, and of Ø 11 µm and 17 µm, finding that SiO_2, Al_2O_3, Fe_2O_3 y CaO represent el 81.17% and 88.84%, with SiO_2 being 51.82% and 50.90% respectively. In this regard [39] indicates that a SiO_2 content higher than 42% determines that the rock belongs to an alkaline type basalt.

3.2 Slump
Figure 3 shows the influence of the amount of BF on the slump of concrete with type I cement for three types of mixes BF-0.10, BF-0.30 y BF-0.60, prepared with Ø 16 µm fibers, length 24 mm. In it, it is seen that the slump decreases when the amount of BF increases, reaching values lower than 3.10", 2.30" and 1.50" for the mixtures BF-0.10, BF-0.30 and BF-0.60 that represent 11.43%, 34.29% and 57.14% less compared to the BF-0 mixture (standard). [40] and [41] studied mixtures with BF for additions with 0.1%, 0.3% and 0.4%, 18 mm in length and Ø 17 µm, and with 0.05%, 0.10%, 0.15% and 0.20%, length 18 mm and Ø 15 µm; finding a decrease of 15.25%, 20.00% and 47.46%, and of 29.71%, 49.43%, 44.86% and 60.71% respectively. [42] indicates that this behavior is due to the increase in the coefficient of friction between the cement and the fiber during mixing and the moisture contained in the fiber, causing a decrease in workability.

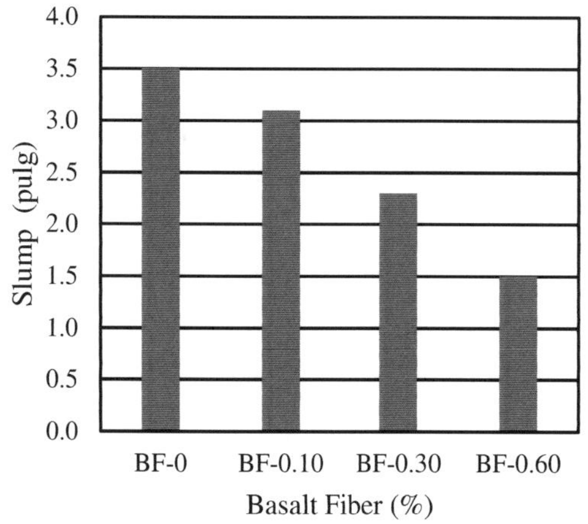

Figure 3. Influence of the amount of BF on the slump.

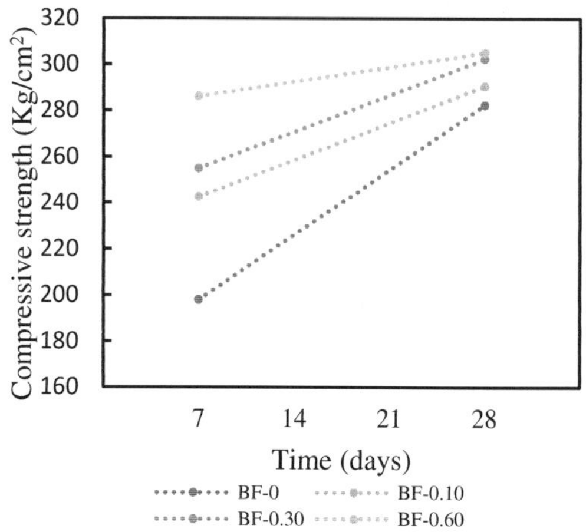

Figure 4. Influence of the test age on the compressive strength of concrete cured in tap water.

3.3 Compressive Strength of specimens cured in tap water

Figure 4 shows the influence of the BF test age on the compressive strength of 3 concrete mixes BF-0.10, BF-0.30 and BF-0.60, prepared with fibers of length 24 mm and Ø 16 μm and cured tap water. It can be seen that the resistance increases for 7 and 28 days, reaching values greater than 286.08 kg/cm^2 and 305.04 kg/ cm^2 for the BF-0.60 mixture, which represent 44.48% and 8.01% more than the BF-0 sample (standard). [43] and [44] studied mixtures with 0.10%, 0.25%, 0.40% and 0.50% addition, 13 mm in length and Ø 12 μm, and with 0.50%, 0.75% and 1.00%, length 24 mm and Ø 16 μm; finding an increase in resistance at 7 and 28 days of 15.97% and 13.15% and of 22.41% and 26.79% respectively. [45] indicates that this behavior may be due to the presence of sufficient moisture to continue hydration of the cement.

3.4 Compressive Strength of specimens cured in sea water

Figure 5 shows the influence of the BF test age on the compressive strength of 3 concrete mixes BF–0.10, BF–0.30 and BF-0.60, prepared with fibers of length 24 mm and Ø 16 μm and cured in sea water. It can be seen that resistance increases for 7 and 28 days, reaching values greater than 281.20 kg/cm^2 and 310.81 kg/cm^2 for the BF-0.60 mixture, which represent 69.80% and 13.00% more than the BF-0 sample (standard). [46] and [47] studied mixtures with 0.10%, 0.20% and 0.30% addition, 20 mm in length and Ø 15 μm, and with 0.15%, 0.20% and 0.25%, length 35 mm and Ø 13 μm; finding an increase in resistance at 7 and 28 days of 8.43% and 4.76% and 44.68% and 7.03% respectively. [48] indicates that this behavior is due to the development of hydration products to block the interior pores of concrete.

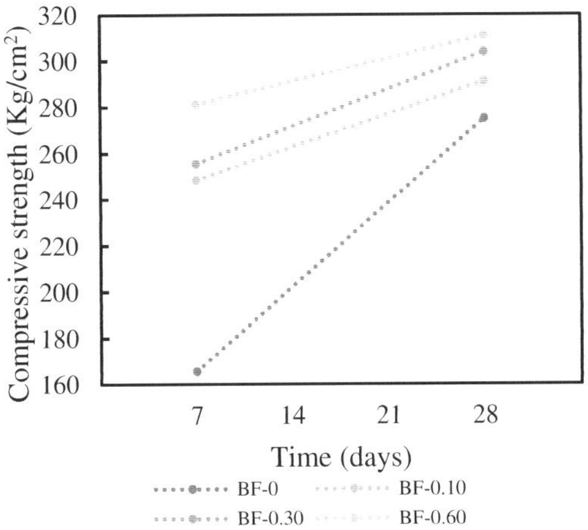

Figure 5. Influence of the test age on the compressive strength of concrete cured in sea water.

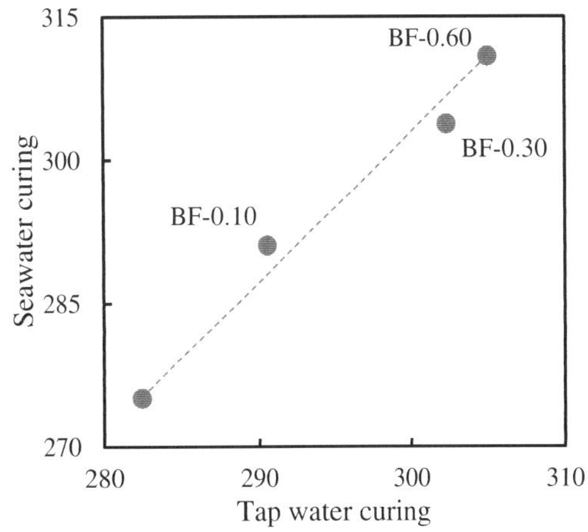

Figure 6. Relationship of curing in sea water vs curing in tap water.

3.5 Relationship of curing with sea water vs curing with tap water

Figure 6 shows the relationship of seawater cure to potable water cure of concrete specimens with BF. In it, it can be seen that the cylindrical 150 x 300 mm specimens cured in seawater have a greater linear relationship and very similar with slightly higher values in 0.17%, 0.45% and 1.89% for BF-0.10, BF-0.30 and BF0.60 that represent an average of 0.84% more. [49] and [50] studied 150x150mm cubic specimens made with fly ash with additions of 25%, 30%, 45%, and 100x200mm cylindrical specimens with 44Kg/m^{3}silicate fume; finding an increase of 6.98% and 0.35% respectively. [51] indicates that this behavior is due to the entry of chloride, sulfate and sodium into the microstructure of the cement paste, clogging the pores and decreasing porosity.

3.6 Carbonation

Table 3 shows the influence of BF on the carbonation depth of concrete with type I cement for a ratio a/c = 0.47 and an exposure time of 90 days. In it, it is seen that by increasing the amount of fiber the carbonation depth decreases, reaching the average values of 150, 136 and 125mm for the addition percentages of 0.10%, 0.30% and 0.60%, which means a decrease in the depth of the 16.67% for BF-0.60 compared to BF-0.10. Figure 7, Figure 8 and Figure 9 show the different carbonation zones for the samples tested. [52] and [53] studied mixtures with addition of 20%, a/c = 0.50 and exposure time of 42 days, and with 25% and 50%, a/c = 0.46 and exposure time of 70 days, finding a decrease in carbonation depth of 40% and 38.88% 11.11% respectively. [54] indicates that this behavior may be related to the lower water-cement ratio of the mix, which influences the uniformity of the porosity of the concrete.

Table 3. Influence of the amount of BF on carbonation.

Identification	Zone	Carbonation Depth d_k(mm)		
		Mínimal	Maximun	Average
BF-0.10	Colorless Purple-red	150	150	150
BF-0.30	Colorless Purple-red	135	138	136
BF-0.60	Purple-red	125	126	125

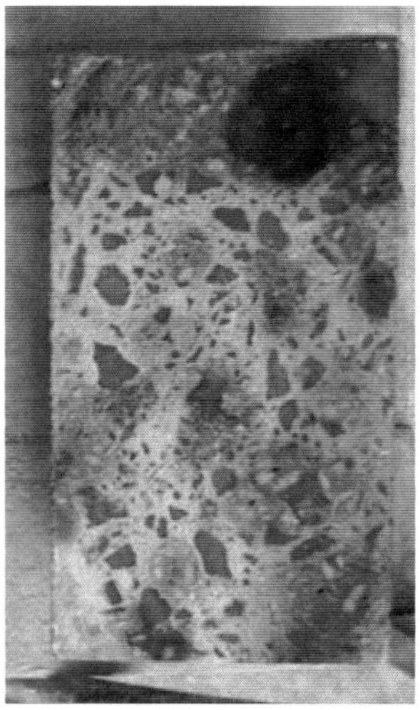

Figure 7. Specimen with 0.10% BF.

Figure 8. Specimen with 0.30% BF.

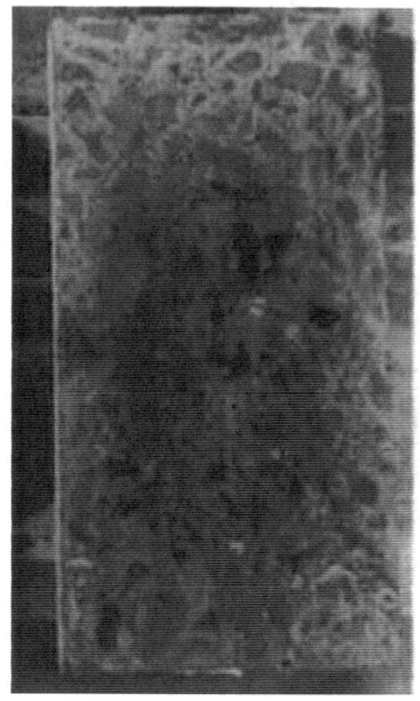

Figure 9. Specimen with 0.60% BF.

4. Conclusions

With the incorporation of the basalt fiber and Type I cement, the formation of cracks in the surface of the piles is delayed because the fibers improve the brittle behavior of the concrete.

The higher composition of SIO_2 in basalt fiber provides better stability under alkaline conditions. The slump when decreasing produces a drier consistency of the mixture, generating a greater difficulty for mixing.

The compressive strength of specimens cured in tap water is increased by constant humidity, vapor retention and protection from contaminating agents.

The higher compressive strength of sea water cured concrete specimens is due to the presence of calcium chloride in seawater, which accelerates the hardening process of the concrete.

The slightly higher relationship between sea water cure over tap water cure is due to the fact that sea salts slightly increase the density of concrete.

The decreasing carbonation depth indicates that the fiber improves the micro-pore structure of the concrete, delaying the penetration of sea water.

5. References

[1] Reyes Díaz E, 2012 *Resistencia A La Corrosión De Concretos Aligerados Con Mezclas Ternarias* Tesis Doctoral CIMAV (Chihuahua, México) p 116

[2] Gambhir M L 2013 *Concrete technology: theory and practice* (New Delhi, India: Tata McGraw-Hill Education) p 774

[3] Zhou H, et al. 2020 Experimental Study on Basic Mechanical Properties of Basalt Fiber Reinforced Concrete *Materials* **13(6)** 1362 1 - 20

[4] Garay L, 2007 *El Ciclo de evolución del destino turístico una aproximación al desarrollo histórico del turismo en Cataluña* Tesis Doctoral UAB (Barcelona, España) p 430

[5] Yu S, & Jin H, 2020 Modeling of the corrosion-induced crack in concrete contained transverse crack subject to chloride ion penetration *Const and Building Materials* **258** 1 - 15

[6] Courard L, et al. 2012 Pathologies of concrete in Saint-Vincent Neo-Byzantine Church and Pauchot reinforced artificial stone *Const and Build Materials* **34** 201 - 210

[7] Yuan M, Xiao M, and Li L 2012 Cracks cause analysis and research on the crane rail beam of Zhanjiang plant coal wharf port *Procedia Engineering* **31** 520 - 527

[8] Monaldo E, Nerilli F and Vairo G 2019 Basalt-based fiber-reinforced materials and structural applications in civil engineering *Comp Structures* **214** 246 - 263

[9] Dhand V, Mittal G, Rhee K, Park S and Hui D 2015 A short review on basalt fiber reinforced polymer composites *Composites Part B: Engineering* **73** 166 - 180
[10] Van de Velde K, Kiekens P and Van Langenhove L 2003 Basalt fibres as reinforcement for composites *Conference on Composites/Nano Engineering* **2026** 20 - 26
[11] Shi F J 2012 A Study on Structure and Properties of Basalt Fiber *Applied Mechanics and Materials* **238** 17 - 21
[12] Jun W, et al. 2010 Experimental research on mechanical and working properties of non-dipping chopped basalt fiber reinforced concrete *Con Inf Man, Inn Man & Ind Eng* **4** 635 - 637
[13] Ma J, Qiu X, Cheng L and Wang Y 2011 Experimental research on the fundamental mechanical properties of presoaked basalt fiber concrete *Advances in FRP Comp in Civil Eng* 85 - 88
[14] Kizilkanat A, et al. 2015 Mechanical properties and fracture behavior of basalt and glass fiber reinforced concrete: An experimental study *Const and Building Materials* **100** 218 - 224
[15] Rathod N, Gonbare M and Pujari M 2013 Basalt fiber reinforced concrete *Int Journal of Science and Research* **14(5)** *359* - 361
[16] Chen F 2013 An experimental study on mechanical properties of basalt fiber reinforced concrete *Applied Mechanics and Materials* **405** 2767 - 2770
[17] Zhang P, et al. 2020 Experimental study on the interfacial bond behavior of FRP plate-high-strength concrete under seawater immersion *Const and Building Materials* **259** 1 - 18
[18] Dawood E T and Ramli M 2012 Durability of high strength flowing concrete with hybrid fibres *Const and Building Materials* **35** 521 - 530.
[19] Olutoge F A & Amusan G M 2014 The effect of sea water on compressive strength of concrete *Int Journal of Engineering Science Invention* **3(7)** 23 - 31
[20] Susilorini M R and Wibowo T 2005 The performance of early-age concrete with seawater curing *Journal of coastal development* **8(2)** 89 - 95
[21] Adiwijaya, et al. 2017 Influence of seawater mixing and curing on strength characteristics and porosity of ground granulated blast-furnace slag concrete *Mat Sci & Eng* **271(1)** 1 - 7
[22] Kaushik S K and Islam S 1995 Suitability of sea water for mixing structural concrete exposed to a marine environment *Cement and Concrete Composites* **17(3)** 177 - 185
[23] Mastali M, et al. 2020 Towards sustainable bricks made with fiber-reinforced alkali-activated desulfurization slag mortars incorporating furnace aggregates *Const & Bui Mat* **232** 1 - 14
[24] Zhao H, Wang L, Wang D, Bai J and Fan Y 2020 Effect of basalt fiber on chloride ion penetration of Reactive Powder Concrete *Materials Science and Eng* **711(1)** 1 - 6
[25] Kirthika S K, Singh S K and Surya M 2018 Durability studies on basalt fiber reinforced concrete *Indian Concrete* **92(4)** 45 - 55
[26] Neville A and Brooks J 2010 *Concrete Technology* (Harlow, England: Prentice Hall) p 442
[27] American Society for Testing and Materials *ASTM C1157 Standard Performance Specification for Hydraulic Cement* 2008
[28] American Society for Testing and Materials *ASTM C33M Standard Specification for Concrete Aggregates* 2018
[29] American Society for Testing and Materials *ASTM C1602M Standard Specification for Mixing Water Used in the Production of Hydraulic Cement Concrete* 2018
[30] Committee 211 ACI 211.1 *Standard practice for selecting proportions for Normal-Heavy weight and mass concrete* 2009
[31] American Society for Testing and Materials *ASTM C192M Standard Practice for Making and Curing Concrete Test Specimens in the Laboratory* 2019
[32] American Society for Testing and Materials *ASTM C1074 Standard Practice for Estimating Concrete Strength by the Maturity Method* 2019
[33] American Society for Testing and Materials *ASTM E1621 Standard Guide for Elemental Analysis by Wavelength Dispersive X-Ray Fluorescence Spectrometry* 2013
[34] American Society for Testing and Materials *ASTM C143M Standard Test Method for Slump of Hydraulic-Cement Concrete* 2020
[35] American Society for Testing and Materials *ASTM C39M Standard Test Method for Compressive Strength of Cylindrical Concrete Specimens* 2020

[36] Asociación Española de Normalización y Certificación, UNE 112 *Corrosión de armaduras. Determinación de profundidad de carbonatación en hormigones endurecidos, en servicio* 2011
[37] Li R, et al. 2017 Radiation shelding property of structural polymer composite: continuous basalt fiber reinforced epoxy matrix composite containing erbium oxide *Comp Sci & Tec* **143** 67 - 74
[38] Quagliarini E, Monni F, Lenci S and Bondioli F 2012 Tensile characterization of basalt fiber rods and ropes: A first contribution *Const and Build Materials* **34** 372 - 380
[39] Militky J and Kovacic V 1996 Ultimate mechanical properties of basalt filaments *Textile Research Journal* **66(4)** 225 - 229
[40] Algin Z and Ozen M 2018 The properties of chopped basalt fiber reinforced self-compacting concrete *Const and Building Materials* **186** 678 - 685
[41] Niu D, et al. 2020 Experimental study on mechanical properties and durability of basalt fiber reinforced coral aggregate concrete *Const and Building Materials* **237** 1 - 12
[42] Wang X, et al. 2019 The Effects of Fiber Length and Volume on Material Properties and Crack Resistance of Basalt Fiber Reinforced Concrete *Advs in Mat Science & Eng* 1 - 17
[43] Brik V B 1999 Performance evaluation of basalt fibers and composite rebars as concrete reinforcement *Tech Res submitted to NCHRP-IDEA* 1 - 97
[44] Kirthika S K and Singh S K 2018 Experimental Investigations on Basalt Fiber-Reinforced Concrete *Journal of The Institution of Engineers (India): Series A* **99(4)** 661 - 670.
[45] Safiuddin M, et al. 2007 Effect of different curing methods on the properties of microsilica concrete *Australian Journal of Basic and Applied Sciences* **1(2)** 87 - 95
[46] Singaravadivelan R, et al. 2012 An experimental study has been carried to evaluate the behavior of concrete elements using basalt chopped strands fiber *Chem, Bio-Chem & Env Sci* 14 - 15
[47] Li M, Gong F and Wu Z 2020 Study on mechanical properties of alkali-resistant basalt fiber reinforced concrete *Construction and Building Materials* **245** 1 - 9
[48] Islam M, M Islam M S, Al-Amin M and Islam M M 2012 Suitability of sea water on curing and compressive strength of structural concrete *Journal of Civil Engineering* **40(1)** 37 - 45
[49] Alaejos P and Bermudez M A 2011 Influence of seawater curing in standard and high-strength submerged concrete *Journal of materials in civil engineering* **23(6)** 915 - 920
[50] Kumar G R and Kesavan V 2020 A review analysis of cement concrete strength using sea water *Materials Today: Proceedings* **22(3)** 983 - 986
[51] Lorenzo M, et al. 2003 Role of aluminous component of fly ash on the durability of Portland cement-fly ash pastes in marine environment *Waste management* **23(8)** 785 - 792
[52] Huang Q, et al. 2015 The influence of fiber on the resistance to chloride-ion penetration of concrete under the environment of carbonation *Material Science and Applications* 223 - 228
[53] Sisomphon K and Franke L 2007 Carbonation rates of concretes containing high volume of pozzolanic materials *Cement and Concrete Research* **37(12)** 1647 - 1653
[54] Wang X, et al. 2020 Carbonation of loaded RC elements made of different concrete types: Accelerated testing and future predictions *Const & Building Materials* **243** 1 - 14

IOP Conf. Series: Materials Science and Engineering 1054 (2021) 012009 doi:10.1088/1757-899X/1054/1/012009

The Physical Quality Assessment of Residential Area in Jabodetabek – Indonesia with Green and Livable Settlement Concept

N Nurdiani[1*], **W Katarina**[1]

[1] Architecture Department, Faculty of Engineering, Bina Nusantara University, Jakarta, Indonesia 11480

*nnurdiani@binus.edu

Abstract. Green architecture as one of the concepts of Green & Livable Settlement has begun to be applied on the designing of residential areas in Jabodetabek (Jakarta, Bogor, Depok, Tangerang, Bekasi). To find out whether the residential area meets the criteria specified in the Green & Livable Settlement concept, it is necessary to conduct a study on the Physical Quality Assessment of Residential Areas in Jabodetabek with the Green & Livable Settlement Concept. The study was conducted using descriptive methods, through surveys, observations and interviews. The results of the study can provide an assessment overview of the physical quality of housing in Jabodetabek. This assessment can be used to improve the architectural design of residential area. The assessment also shows that the use of the green theme gives the benefit for the promotion of property, and the physical quality of residential area better than other residential area.

1. Introduction

At present most of the world's population lives in urban areas, and will continue to increase to almost around 70-80%. Around 20-30% growth of cities in developing countries will quickly match the percentage of city dwellers in developed countries [1]. This population development requires a new residential area that is not entirely available in urban areas. The development of new residential areas on the borders of the city of Jakarta and surrounding areas, often called Jabodetabek (Jakarta, Bogor, Depok, Tangerang, and Bekasi), requires a new concept that is in line with current needs, namely the design of residential areas that are part of environmental conservation efforts. Because of this need, the concept of Green & Livable Settlement began to be applied in new residential areas.

Architecture plays a role in the development process that occurs on earth, because architecture occupies space on natural land. So that architecture can be seen as a cause of environmental degradation or damage to nature in a developing city, through the construction of various buildings for settlements and industries, as well as other supporting facilities that do not pay attention to environmental and ecological conditions. The phenomenon of global warming caused by the greenhouse gas effect on earth is also believed by researchers to be caused by development activities [2].

At the beginning of the concept of sustainable development, design focused on only one issue, namely the efficiency or use of used materials. But since the 1980s and 1990s, planners and designers have begun to realize that the integration of all factors can produce good sustainable designs. So that today, many sustainable buildings have been combined into several focus areas of discussion such as;

land selection, energy efficiency, water conservation, efficiency of material resources, and environmental quality in space [3].

Currently, one of the concepts of Sustainable Architecture, Green Architecture, has begun to be applied not only to residential design, but also to the design of new residential areas with the Green Settlement concept. Even the concept of Livable Settlement is also a concept that is used as more value in the design of residential areas. With the application of the Green & Livable Settlement concept, it is hoped that residents in the area can live more comfortably and better with quality environmental support.

To find out whether the new residential area that applies the Green & Livable Settlement concept has good physical quality according to the criteria that must be met, it is necessary to study the Physical Quality Assessment of Residential Areas in Jabodetabek with the Green & Livable Settlement Concept. The study was conducted in a new residential area in Jabodetabek. The results of the study can provide an overview of the physical quality of the residential area, which can be used to improve the architectural design of the next residential area to be better physical quality.

The sustainable framework has been submitted by the UIA (International Union of Architects) through the Copenhagen Declaration on 7 December 2009. The UIA Declaration states that buildings and the construction industry have an impact on climate change that is happening right now. And these various impacts can be reduced by determining the built environment system. UIA is committed to reducing the impact of damage through a "Sustainable Design Strategy" [4].

A study of the Physical Quality Assessment of Residential Areas in Jabodetabek with the Green & Livable Settlement Concept has never been conducted. Research that has been conducted is an assessment of the application of the concept of green architecture in residential homes in the city of Medan [5] and in Jakarta [6,7]. The results of both studies concluded that not yet fully residential design in the city of Medan and in the city of Jakarta applied all aspects specified in the criteria of green architecture.

2. Research methodology

Research on the Physical Quality Assessment of Residential Areas in Jabodetabek with the Green & Livable Settlement Concept is a descriptive study. The purpose of this study was to identify the architectural design of residential areas using the Green & Livable Settlement concept, and to enrich knowledge about the physical quality of residential areas using the Green & Livable Settlement concept. Data collection in the field is done by field surveys, observations and interviews. Data analysis was performed with a descriptive approach using Greenship Neighborhood version 1.0.

The study took place in Jabodetabek. The research setting of Physical Quality Assessment of Residential Areas in Jabodetabek with the Green & Livable Settlement Concept was carried out on new residential areas that applied the concept in the cities of Jakarta, Bogor, Depok, Tangerang, and Bekasi. The study was conducted from March 2019 to November 2019 (9 months).

3. Results and discussion

The building and green area assessment system in Indonesia was developed by the Green Building Council Indonesia (GBCI). The GBCI, which was founded in 2009, is an independent and non-profit institution committed to implementing the principles of sustainability by developing parameters or benchmarks called the Greenship. The assessment in the application of building practices and environmentally friendly areas can be carried out starting from the planning, construction and maintenance of buildings.

The research to evaluate the application of the green concept by using the greenship parameter took the case study locations in the cities of Jakarta, Bogor, Depok, Tangerang, and Bekasi, which were chosen purposively, namely in residential areas which claimed to apply the Green concept (see figure 1). The locations of selected residential areas are GLC housing in West Jakarta, SC housing in Bogor, TGS housing in Depok, BSDC housing in Tangerang, and LC housing in Bekasi.

IOP Conf. Series: Materials Science and Engineering 1054 (2021) 012009 doi:10.1088/1757-899X/1054/1/012009

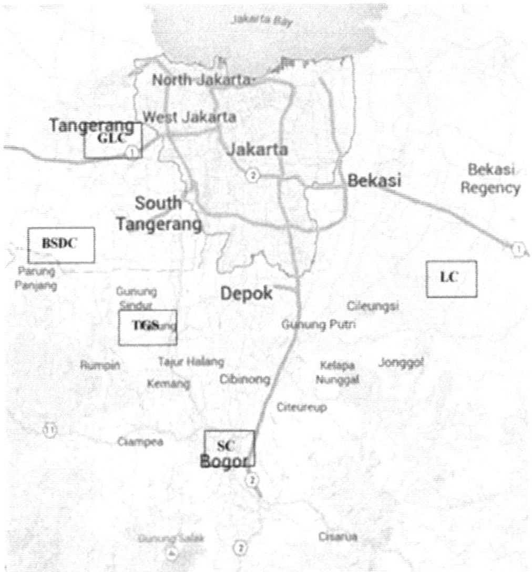

Figure 1. Residential area chosen as a case study in Jabodetabek.

Figure 2. The physical condition of residential areas in Jabodetabek.

These five housing locations were chosen because they have a residential area that is almost the same size, and also has a theme with a Green Insight Residential Area design with the Green Architecture theme approach. The development of residential areas also has the same design model which is divided into residential clusters, where each cluster is equipped with supporting facilities to meet their social and economic needs (see figure 2).

The development of residential type models in clusters in five housing locations has a similarity, that is, the developer designs dwellings as needed with a facade that adopts architectural forms and elements based on foreign architectural characteristics / styles (European, Mediterranean, American, etc.). This condition has become a challenge for Indonesian architects to develop and introduce modern (contemporary) Indonesian architecture in order to be able to host their own country.

Before doing the assessment, it is necessary to analyze the feasibility of the five housing estates. After being considered feasible, an assessment of compliance with the green architecture criteria of the five residential areas was carried out by using Greenship Neighborhood version 1.0. (see figure 3). The assessment results are then adjusted to the Greenship Ranking Level (see table 1).

No	Aspect	Benchmark	Point	GLC	SC	TGS	BSDC	LC
1	Enhancement of land ecology	Green Base Area	P	OK	OK	OK	OK	OK
		Public Green Area (25% - 35%)	4	3	2	2	3	2
		Habitat Preservation	6	2	0	0	2	0
		Land Revitalization	4	0	0	0	0	0
		Micro Climate	3	2	1	1	3	1
		Local Food	2	0	0	0	0	0
		Sub Total	19	7	3	3	8	3
2	Movement and Connectivity	Movement and Connected Areas Study	P	OK	OK	OK	OK	OK
		Pedestrian Path Design Strategy	10	6	4	4	4	4
		Public transportation	6	4	2	0	4	4
		Utilities and Public Facilities	2	2	2	2	2	2
		Universal accessibility	3	2	0	0	0	0
		Bike Networks and Place	3	3	0	1	0	0
		Shared Parking	2	1	1	1	1	1
		Sub Total	26	18	9	8	11	11
3	Water Conservation and Management	Calculation of Water Consumption	P	OK	OK	OK	OK	OK
		Alternative Water Sources	6	2	0	0	0	0
		Rainwater Runoff Management	7	7	4	3	4	3
		Preservation of Water Bodies and Wetlands	2	2	0	0	0	0
		Liquid Waste Management	3	3	0	0	0	0
		Sub Total	18	14	4	3	4	3
4	Solid Waste Material	Operational Solid Waste Management	P	OK	OK	OK	OK	OK
		Advanced Solid Waste Management	4	2	2	2	2	2
		Construction Waste Management	4	3	3	2	3	3
		Road Infrastructure Regional Materials	4	4	4	4	4	4
		Road Infrastructure Recycled Materials	2	0	0	0	0	0
		Sub Total	14	9	9	8	9	9
5	Community Welfare Strategy	Community Facilities	2	2	2	2	2	2
		Social and Economic Benefits	4	4	2	2	4	2
		Community Concern	4	2	1	1	2	1
		Mixed Area	2	2	2	2	2	2
		Local Culture	2	0	0	0	0	0
		Safe Environment	2	2	2	2	2	2
		Sub Total	16	12	9	9	12	9
6	Buildings and Energy	Greenship Green Building	6	0	0	0	2	0
		Balanced Occupancy	1	1	1	1	1	1
		Energy Efficiency in the Region	4	2	2	2	2	2
		Alternative Energy	3	0	0	0	0	0
		Light Pollution Reduction	2	2	2	2	2	2
		Noise Pollution Reduction	2	2	2	2	2	2
		Sub Total	18	7	7	7	9	7
7	Innovation and Innovation Development	GA / GP empowerment for Managers	3	1	1	0	1	1
		Area Management	2	4	2	2	2	2
		Inovation	6	2	0	0	2	1
		Sub Total	11	7	3	2	5	4
		TOTAL	**122**	**74**	**44**	**40**	**57**	**47**
		Ranking	**-**	**I/P**	**4/BtS**	**5/B**	**2/G**	**3/S**

Figure 3. Assessment of the application of green architecture to the residential area.

Table 1. Greenship ranking level points.

Points to be earned	Planning & Design	Built Project
Platinum (P)	56	74
Gold (G)	44	58
Silver (S)	35	47
Bronze (B)	27	35

The results of the simulation assessment for the submission of residential area greenship show that of the five residential areas that claim to apply the concept of green architecture, they have tried to make it happen in the area and residential design. Proven five of them showed the results of the

assessment that made it possible to get a greenship ranking with varied results. The results of the assessment are:
1. GLC housing has the highest rating (platinum) which means that the results of the simulation assessment meet almost all of the standards that must be met from green architecture criteria (74/122 = 60.7%).
2. BSDC housing received a high rating (gold) which means that the results of the simulation assessment met most of the standards that must be met from the green architecture criteria (57/122 = 46.72%).
3. LC housing has a medium rating (silver) which means that the results of the simulation assessment meet some of the standards that must be met from the green architecture criteria (47/122 = 38.5%).
4. SC housing has a sufficient rating (bronze to silver) which means that the results of the simulation assessment will soon reach some of the standards that must be met from green architecture criteria (44/122 = 36.1%).
5. TGS housing received a sufficient rating (bronze) which means that the results of the simulation evaluation were sufficient to meet some of the standards that must be met from the green architecture criteria (40/122 = 32.8%).

The results of the assessment of the application green architecture concept chasing targets at high (gold) and highest (platinum) positions. Of the 5 selected locations, only 2 locations have reached that position, namely GLC and BSDC housing.

4. Conclusion
Assessment of the application of the concept of green architecture in residential areas in Jabodetabek using Greenship Neighborhood version 1.0. of 5 housing locations showing the results that only 2 residential areas (GLC and BSDC) can be considered successful in meeting green architecture criteria. There are opportunities in 3 other residential areas (LC, SC and TGS) to improve themselves in order to meet all the architectural design parameters and green areas based on Greenship Neighborhood version 1.0. The assessment also shows that the use of the green theme for the benefit of the promotion of property developers to buyers has been pursued and applied well by the developer in the design of residential areas and buildings.

5. References
[1] Kuswartojo T 2006 Asas Kota Berkelanjutan dan Penerapannya di Indonesia *Jurnal Teknologi Lingkungan P3TL-BPPT 7* (1) 1-6
[2] Ervianto W I 2012 *Selamatkan Bumi Melalui Konstruksi Hijau* (Yogyakarta: Penerbit Andi)
[3] Itewi M & Alsafseh Q 2012 Green Home with Energy Saving Design – Eco Homes in Jordan *Europen Scientific Journal 8* (21) 107-117
[4] Tanuwidjaja G 2011 Desain Arsitektur Berkelanjutan di Indonesia: Hijau Rumahku Hijau Negeriku *Prosiding Seminar Nasional Lingkungan Hidup, Living Green: Mensinergikan Kehidupan, Mewujudkan Keberlanjutan* (Surabaya: Universitas Kristen Petra)
[5] Alvan S, Novrianty I, Nst Lynna P A & Luthan 2015 Konsep Desain Rumah Tinggal Berkelanjutan di Kota Medan *Jurnal Sosek Pekerjaan Umum 7* (1) 1-79
[6] Nurdiani N, Katarina W, Setiawan D, & Sakina B 2018 Penerapan Konsep Arsitektur Berkelanjutan pada Rumah Tinggal Perkotaan *Prosiding Seminar Nasional Infrastruktur Berkelanjutan* (Malang: Institut Teknologi Nasional)
[7] Nurdiani N, Taufik 2020 The study of application of the green architectural design concept at residential area in Jakarta *IOP Conference Series: Earth and Environmental Science* 426 (2020) 012067

Acknowledgement
Thank you to Bina Nusantara University for providing research funding through BINUS Applied Grants 2019, so that research can be carried out and completed well. Hopefully the research results can be useful for the enrichment of knowledge and for the community.

Thermal Comfort in the Modern Movement – Consequences of Constructive Definition on the Summer Behaviour of a Housing Building in Porto, Portugal

I Silva [1*], P Mendonça[1] and C Maia [1]

1 University of Minho, School of Architecture, Lab2PT, Guimarães, Portugal

*ivo_1284@hotmail.com

Abstract. The knowledge of the properties of construction materials, as well as the way they are organized in the construction systems, namely those of the facades, are essential to understand the functional performance of buildings, namely their thermal comfort. This article presents a research oriented towards the constructive characterization of the facades of a multifamily building and the consequences for its thermal comfort in summer. The building here analysed is Edifício Lino, built between 1951 and 1954 in Porto, designed by the architects Arménio Losa and Cassiano Barbosa, prominent figures in the architecture of the modern movement in Portugal. The importance of studying buildings at this time is related to the need to define strategies for use, maintenance and possible intervention in the existing heritage in order to respond to the contemporary expectations of habitability of these works of the Modern Movement, often forgotten due to the temporal proximity. From the study carried out, it can be concluded that thermal comfort emerges as a central concern of the reference architects of this time, who considered this issue an inexorable architectural quality.

1. Introduction.

"The intervention in modern heritage buildings are not simple, the whole process has to be very careful with the questions of materiality, authenticity, it is an architecture that has a very own spatiality that often does not accept great changes [1].

This article is based on the results of a research where architectonic principles, materiality and construction choices taken by two prominent architects of the Portuguese Modern Movement - Arménio Losa and Cassiano Barbosa, at the Multi-family housing project - Edifício Lino, from 1953, are contrasted with comfort parameters. Through these parameters and computer simulations we pretend to create the fundamental bases of a restoration intervention to be carried out in the future. This intervention shall be attentive to the temporal, spatial and tectonic peculiarities of this kind of construction.

The Modern Movement architecture has always search for the well-being and the comfort of man. Housing, work, body and mind culture [2] are the principles of this architecture.

"The notion of comfort appears as a clearly modern feature. (...) modern comfort is considered as an architectural quality, and it is necessary to attribute it to furniture, certain finishes, efficient mechanization of the kitchen and bathroom, and (..) elements that control the environment." [3]

This notion of incorporating comfort as intrinsic quality of the architecture was always present on all the works of Losa and Barbosa. Prior to any restoration project we must verify how these materials and qualities have behaved over time and, above all, confirm that the comfort values (thermal, light and noise) of these buildings are acceptable for contemporary standards. As Ana Tostões argues "the current regulations on safety, energy performance and environmental comfort put reuse actions in pair with the requirements necessary for the construction of new work, often calling into question a qualified recovery. Thus, one of the objectives of the restoration will have to go through the meeting of a specific lexicon, which considers the character of the built along with the mutations necessary for the experience of space." [4]

With the measurements carried out in situ we will be able to verify how the mechanisms designed in order to respond to the demand for comfort, imposed by the architects of the modern movement correspond to contemporary expectations of habitability and legislation and which changes will have to be implemented. Good solar orientation, window frames, ventilation shading systems, insulation and waterproofing systems were part of the construction lexicon applied in all the works of these authors; these systems were designed for the well-being and comfort of modern man's homes.

The future intervention should be able to decide which strategy to implement that optimizes these systems, and that maintains the identity, whether through changes, replacements of elements or simple reparation. The characterization of all systems and materials that integrate the construction of the apartments of the Edifício Lino is fundamental for a complete evaluation of the type of actions to be taken for its efficient use, assuming that the main objective is to maintain identity and integrity of the building. For this and as shown in figure 1, it is necessary to understand the building from its historical framework to the current state.

Figure 1. Edifício Lino, from implantation to detail.

2. Climate context and thermal characterization of the building

The climatic framework of Porto and consequently of Edifício Lino is important to verify its compliance with the reference Thermal Transmittance coefficients established by Portuguese Ordinance No. 379-A / 2015, required for each portuguese climate zone. The objective is to understand the response of the materials used in the envelope constructive systems of one representative apartment in Edifício Lino, in order to perceive its compliance or not with the values stipulated by portuguese legislation. Table 1 presents the climatic data of the city of Porto.

Table 1. Porto city weather data.

City	Porto
Winter climate zone	I2
Number of degree days (GD) (ºC days)	1610
Heating season duration (months)	6.7
Summer climate zone	V1
Average outdoor temperature (ºC)	15.3
ZREF (m)	9

The Thermal Transmittance (U-value) defines the overall ability of conductive or convective barriers to transmit heat. It characterizes the amount of energy in the form of heat that is transmitted through 1 m² of wall surface, in the period of one second, when the temperature difference between the inside and outside is 1 ºC. The coefficient of thermal transmission of a building element is calculated by dividing the thickness of the material (e (m)) by the value of thermal conductivity (λ (W/m°C)) resulting in the thermal resistance of the building elements. Using the calculated value of the thermal resistance Rt, of a given building element, the Thermal Transmittance, U-value. The calculation of the total thermal resistance of a given building element is calculated by Equation 1 as shown in EN ISO 6946:2007.

$$R_t = R_{si} + \sum R_i + R_{se} \ (m^2 .\ ^\circ C/W) \quad (1)$$

where:

Rt - Total thermal resistance of the building element;
Rsi - Interior Surface thermal resistance;
Ri - Thermal resistance of the building element layers I;
Rse - Exterior Surface Thermal Resistance.

It should be noted that ITE50 of the LNEC (National Laboratory of Civil Engineering) [8] indicates values for Ri ,Rsi and Rse for the present thesis used the values of Rse 0.04 and Rsi of 0.13, defined for the horizontal heat flow, due to the fact that all the envelope elements with thermal requirements are vertical so the flow is horizontal. The thermal resistance value of the non-ventilated air gap (that exists inside the walls) was also considered, being this 0.18 (m². ºC/W).

The U-value can be obtained from the thermal resistance according to Equation 2:

$$U = 1/ R_t \ (W/m^2 .\ ^\circ C) \quad (2)$$

The U value of the 4mm plain glasses used in Edifício Lino with wooden frames were also retrieved from ITE50, being this 5.1 (W/m² ºC). Table 2 presents the reference Thermal Transmittance of envelope elements.

Table 2. Reference Thermal Transmittance of opaque and glazing elements, Uref [W/(m².°C)] in accordance with Ordinance No. 379-A/2015.

Uref [(W/ m² ºC)] Current zone of envelope		Climatic zone I2
In contact with the outsider or with non-useful spaces with Heat loss coefficient btr>0.7	Vertical opaque elements	0.40
	Horizontal opaque elements	0.35
In contact with other buildings or non-useful spaces with Heat loss coefficient btr 0.7	Vertical opaque elements	0.70
	Horizontal opaque elements	0.60
Glazed spans (doors and windows) (Uw)		2.40
Elements in contact with the ground		0.50

Thermal inertia allows to dampen temperature variations, keeping it more stable inside the building, therefore further away from extreme values causing discomfort. Thermal inertia can be calculated using the Equation 3:

$$It = \Sigma M.si.Si\ /Ap \quad (3)$$

where :
Msi-Useful surface mass of element i [kg/m²];
Si-Area of the interior surface of element i [m²];
Ap- Useful area of the Pavement [m²].

According to the portuguese regulation (Decree Law n.º 118/2013) three classes of thermal inertia are defined: strong, weak and medium. The Edifício Lino Apartment in study has a strong thermal inertia class due to its high thermal mass, It>400 kg/m², as it can be seen in Table 3.

Table 3. Thermal inertia calculations of the Apartment in study.

Elements	Si (m²)	Ms*i* (kg/m²)	r	Si. Msi. r
Exterior wall	43.5	130	0.5	2828.2
Interior wall type 1	20.4	150	1	3182.4
Interior wall type 2	38.5	198	1	7619.1
Interior wall type 3	19.6	240	1	4704.0
Roof	57.8	150	0.5	4336.5
Pavement Slab type 1	43.8	150	0.5	3285.0
Pavement Slab type 2	14.2	150	0.5	1065.0
Totals	58 m² Ap	1174 kg/m² Ms*i*		
Thermal Inertia Si. Msi. r /Ap				**465.86 kg/m²**

The definition of the envelope elements help understand which walls present thermal requirements and thus realize if the organization of the different materials in the construction systems are able to comply with the regulations in terms of Thermal Transmittance. The analysis of figures 2 and 3 show the type of envelope elements in which the Thermal Transmittance should be considered: the walls of the north and south facades (in red), the respective glazings because they are in contact with the exterior and the wall between the apartment and the common staircases of the building (in yellow). This is because it is a space in direct contact with outside, with fixed windows. It should be noted that the envelope elements marked in green, when in contact with spaces that are also interior, are spaces without thermal requirements. It is important to emphasize that in the walls between apartments, the architects kept the wall with two rows of bricks and air gap between them, for sound insulation; while on the east wall of the building that is in contact with another collective housing, without thermal requirements, they only applied a row of bricks, demonstrating sensitivity to an efficient resources' management.

Figure 2. Plan with envelope elements.

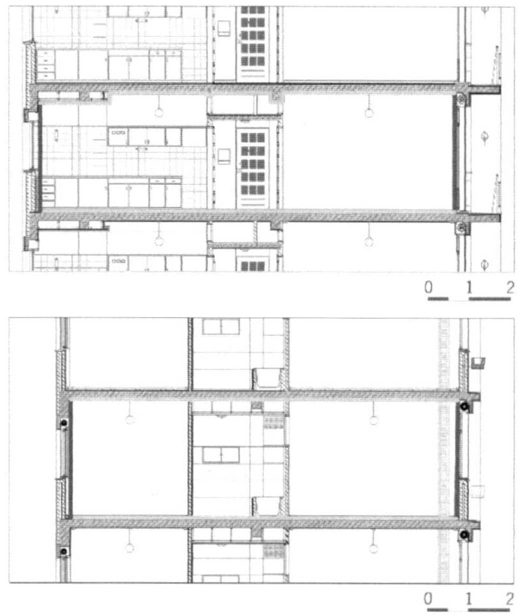

Figure 3. Section [AA´] (top) and Section [BB`] respectively with envelope elements.

The constructive details of the north and south facades were characterized, as well as the wall between the apartment and the staircase in order to calculate the value of the thermal transmittance (U [W/m² °C]) of each of these elements. This calculation allowed to understand whether the elements with thermal needs are in accordance with current standards. The constructive details presented in figure 4 are redesigns of the originals elaborated by the architects.

2.1 Analysis of results

The calculation made and presented to the different building elements of the apartment under study, reveals that none of these meets the thermal requirements of Ordinance No. 379-A/2015 presented in Table 2. The north facade where the kitchen and the north bedroom are located, and the south facade with the living room and the south bedroom, being vertical opaque elements have as U_{ref} the value of 0.40 (W/ m² °C), however in the present building, with all these elements exceeding 1 (W/ m² °C) these requirements are not met. The same applies to elements in contact with non-useful spaces such as the dividing wall between the apartment and the vertical access that exceed the U_{ref} value of 0.70 (W/m² °C) for this type of elements, i.e. also do not meet this requirement.

The glazed spans also do not meet, given the value taken from the ITE50 for simple 4mm glazed spans with wooden frames, of 5.1 (W/m² °C) to be much higher than the U_{ref} requirement of 2.40 (W/m² °C). The weighted mean, in the relationship between the area and the building elements increases the U values, accentuating the discrepancy between the recommended and the calculated values.

The study of the vertical sections allows us to realize that the architects of the work had a logic in the distribution of the chosen materials, the order of the ceramic bricks following from the outside to the interior is: brick of 6cm thick, then the air box of 5 cm and then 8 cm of ceramic brick as the last material i.e., there could be here the intention of the latter , with greater thickness, allow to increase the capacity of thermal accumulation, although slightly there is a clear intention here to obtain better thermic and acoustic comfort inside.

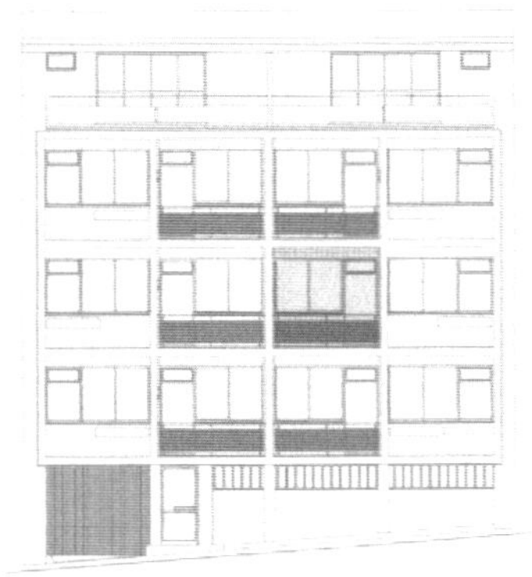

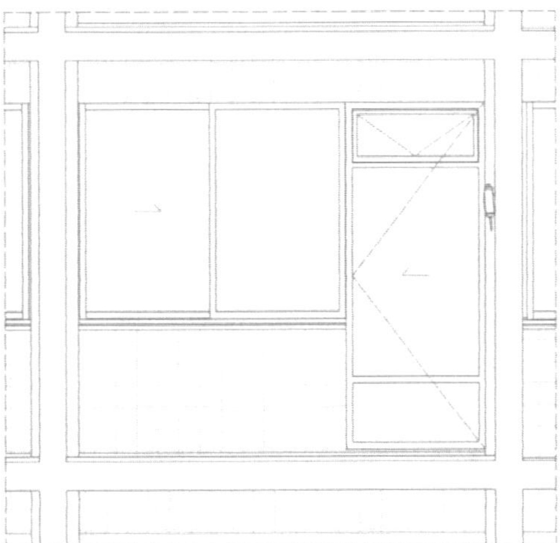

Living Room facade 1:50

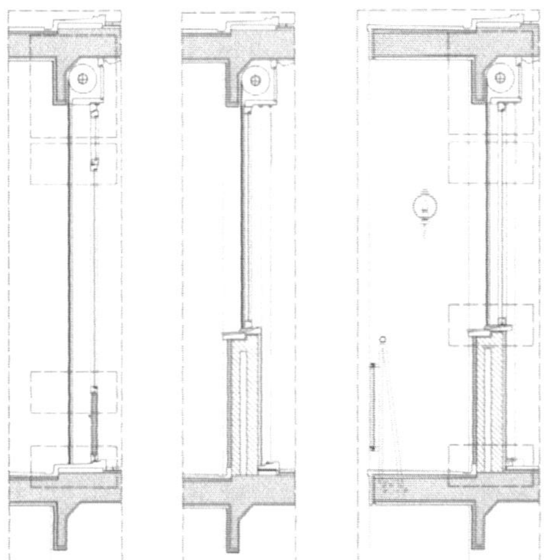

Vertical Sections 1:50

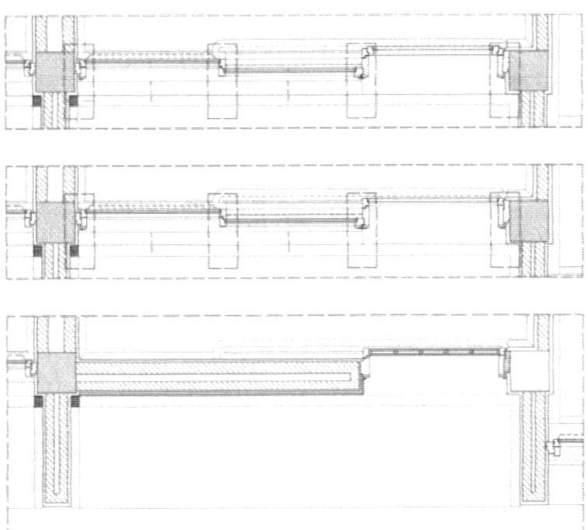

Horizontal Sections 1:50

Building Material (indoor exterior)	e (m)	λ (W/m°C)	Rt [m².°C/W]
Tile	0.006	1.30	0.005
Mortar lime and sand	0.02	0.80	0.03
Ceramic brick 6	0.06	0.34	0.18
Air box	0.05		0.18*
Ceramic brick 8	0.08	0.34	0.24
Plaster	0.02	0.80	0.03
Stucco	0.01	0.40	0.00
Total e (m)	0,237	Σ Rt	0,649
Interior surface thermal resistance, Rsi [m².°C/W]			0,13
Exterior surface thermal resistance, Rse [m².°C/W]			0,04
Thermal transmission coefficient of wall U (W/ m² °C)			1,22
Thermal transmission coefficient Simple glass 4mm with wooden frames U (W/ m² °C)			5.1*
Weighted Average of the value of U (W/ m² °C)			3,36

*Value tabled by ITE50

Thermal transmission coefficient calculation

Figure 4. Constructive details of the living room oriented to the south and respective calculation of the thermal transmission coefficient of the facade.

3. Thermal comfort evaluation - "In situ" Measurements in cooling season

After the results obtained in the constructive analysis and its calculation of the U, realizing that these doesn't correspond to the minimum required values, comfort measurements were carried out on site with the expectation that these may demonstrate that the living conditions of this building are good or at least acceptable. These measurements also allowed the validation of the conditions considered in the thermal simulations that were intended to be carried out.

The experimental evaluation of the hygrothermal performance of the building required monitoring during the cooling season in order to obtain key summer data to understand the thermal behavior of the building. Being an ongoing investigation there is still the intention of conducting the same type of monitoring in winter.

Measurements of temperature and relative humidity were performed in situ by placing portable temperature and humidity monitoring data-logger equipment, two indoors and one on the outside that record the desired values for the necessary parameters. These data needed to reach the operative temperature, important for the study.

For the measurements, two scenarios were used, one with all blinds closed (Scenario 1) and the other with all blinds open (Scenario 2). During the tests performed there were no changes in the opening of the glazed windows, which were always closed. It is also important to mention that the apartments were without any occupancy and therefore without internal gains.

The choice of scenarios is due to the approximation of the usual conditions of unoccupied buildings, also allowing and in order to optimize the intervention, develop and verify the values of comfort in an ideal scenario of simulated occupation.

The results presented in this study were obtained in a measurement campaign carried out in the building, in a period of seven days, representative of the cooling station - May 21 to 29, 2020. Although this period doesn't correspond exactly to summer, it was an atypical week of spring with very high temperatures that corresponded to what was expected from a summer campaign.

These measurements highlight in Scenario 1 the maximum and minimum temperatures in this period, which inside the apartment were 24.8°C and 19.3°C, and outside, 34.8°C and 15.3°C, demonstrating that, while the difference of the indoor temperature peaks is 5.5°C, outside this value is 19.5°C. Comparing the temperature peaks of the interior with the outside, the difference between the maximum temperature of the outside and the inside is 10°C and between the minimums if of 4°C.

Although these results in the apartment are favorable, they are not in line with the luminous comfort, because there is no entry of natural light into the spaces. In Scenario 2 the indoor temperatures present a slightly higher variation, of 7°C, and the maximum temperature in this period was 27.3°C, and a minimum of 20.3°C. Comparing the tests in the difference between the peaks, although smaller than in Scenario 1, are relevant due to solar gains during the day and heat losses through the openings at night. This difference between temperatures is 7.5°C and 5°C in relation to the maximum and minimum peak respectively. This proved that the studies and solar concerns of the architects in the building design, has positive results in the summer season. The results of this process allowed us to understand how the design of the building responds in this solstice where the angle of solar height is greater and, as it is possible to observe in figure 5, with the angle of 70° used by Losa and Cassiano for the solar study in its buildings, the Lino Building is no exception and allows natural lighting during summer in the spaces, but mostly without direct incidence, which would consequently increase the interior temperature.

For Relative Humidity, in both Scenarios they met EN 15251:2007, which recommends for existing Category III buildings, a maximum and minimum value between 70 and 20%. In Scenario 2, the building reaches category II i.e. for rehabilitated buildings. The justification of these positive values is due to the high temperatures recorded outside that provide lower humidity, which consequently, inside the apartment, without any sun protection of the blinds, allowed the indoor temperature to increase and thereby reduce the relative humidity. It can be concluded that by the correct use of blinds, during this season, the values can even be optimized and provide the best comfort, whether thermal or luminous.

"Temperature and humidity are the most important aspects of indoor environmental conditions (...) determine to a large extent the conditions of thermal comfort, due to the impact they have on several of the thermoregulation mechanisms of the human body" [5]

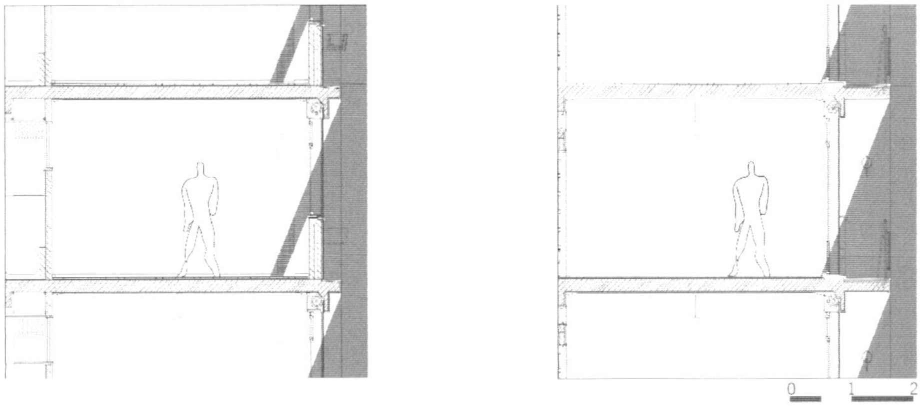

Figure 5. Solar incidence, angle of 70º in summer. South Room left. Living Room right.

To better compare the hygrothermal comfort in the apartments, the Operative Temperature were thus evaluated during the cooling periods, with Scenarios 1 and 2 indicated above. This evaluation also allowed to understand the comfort conditions defined by EN 16798:2019 in order to understand whether the apartments are within the parameters for category III (Pre-existing building category). Graphs shown on figures 6 and 8 were plotted from the operative temperature for an occupant sitting in the center of the compartment in summer clothes (0.5 clo). The indoor air velocity was considered as zero due to all the windows being closed and the average radiant temperature equal to air temperature.

For these graphs, the temperature and maximum and minimum relative humidity of each day were evaluated, in order to establish an average weekly value of each factor, so that through the online tool CBE (Center for the Built Environment) Thermal Comfort Tool, it was possible to verify that the values collected in the building are in accordance with the comfort zones of the EN-16798 standard.
Based on the interpretation of the Graphic presented in figure 6, it is observed that it doesn't comply with the EN-16798 standard in Scenario 1. However, this only happens due to the inoccupation of the building - although the collected values are outside the comfort range of the standard, it should be noted that the temperatures are lower, i.e. it meets the norm by excess, with no cooling needs. Easily in this season and conditions one could reach the values of the norm, just by opening the blinds, which was done in Scenario 2.

IOP Conf. Series: Materials Science and Engineering 1054 (2021) 012010 doi:10.1088/1757-899X/1054/1/012010

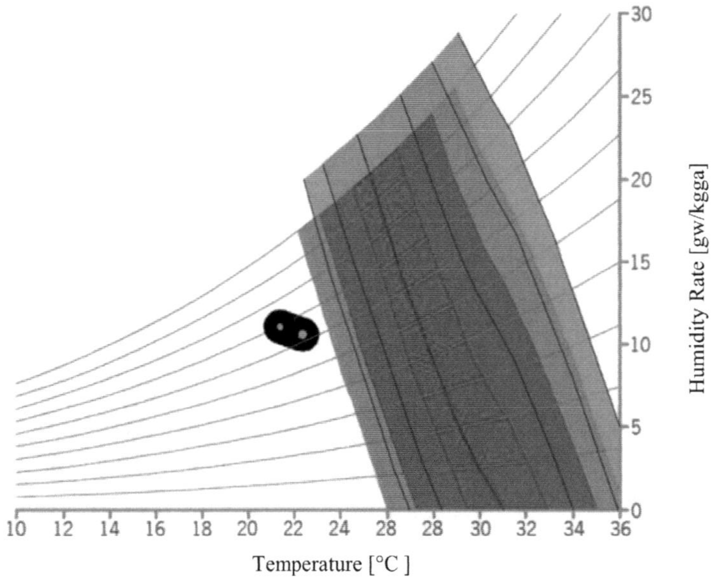

Data entered for maximum temperature values:

Air Temperature: 22,4°C
Radiant Average Temperature: 22,4°C
Air speed: 0 m/s
Relative Humidity: 62,7%
Metabolic Rate: 1 met
Clothing Level: 0,5 clo

Data entered for minimum temperature values:

Air Temperature: 21,4°C
Radiant Average Temperature: 21,4°C
Air speed: 0 m/s
Relative Humidity: 69,6%
Metabolic Rate: 1 met
Clothing Level: 0,5 clo

Legend:

Maximum average temperature recorded
Category I Category II Category III

Minimum average temperature recorded
Category I Category II Category III

Results of the apartment with closed blinds

Figure 6. Evaluation of comfort in the apartment with blinds closed, from 21 to 29 May.

In Scenario 2 the operative temperature has a greater variation throughout the day, due to the opening of blinds, since with the simple glazed of 4mm there are more solar gains during the day and consequent increase in the temperature of this surface and heat losses at night with the decrease of the temperature.

In figure 7 it is possible to observe the greatest variation of the exterior temperature in relation to the interior operative temperature. This reveals the importance of thermal mass, because the thermal discomfort is due to these high variations if these are not absorbed.

Through the consultation of Graphic presented in figure 8 allows verifying compliance with the EN-16798 standard, throughout the week, between category III when it reaches the lowest operative temperature and category I on the days where they are higher, it is important to emphasize that the building, despite having the blinds open throughout the day, always remains within the comfort zone recommended by the standard.

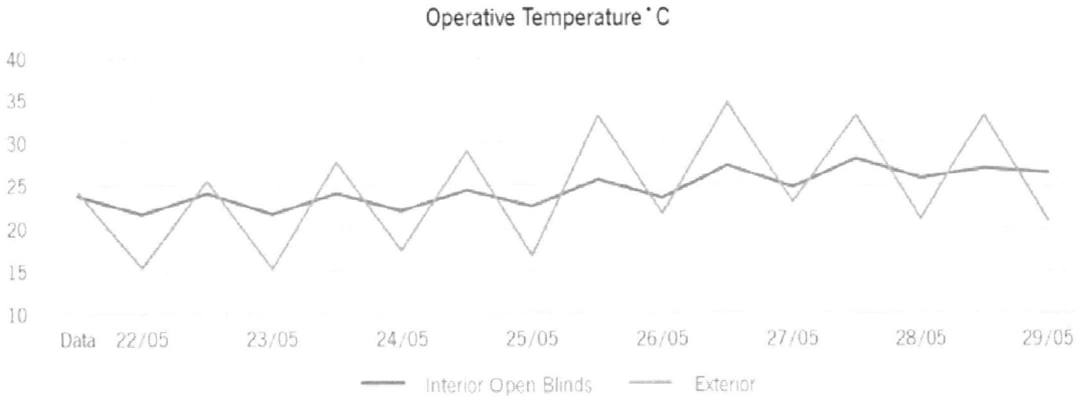

Figure 7. Operative temperature and exterior temperature from May 21 to May 29. Scenario 2.

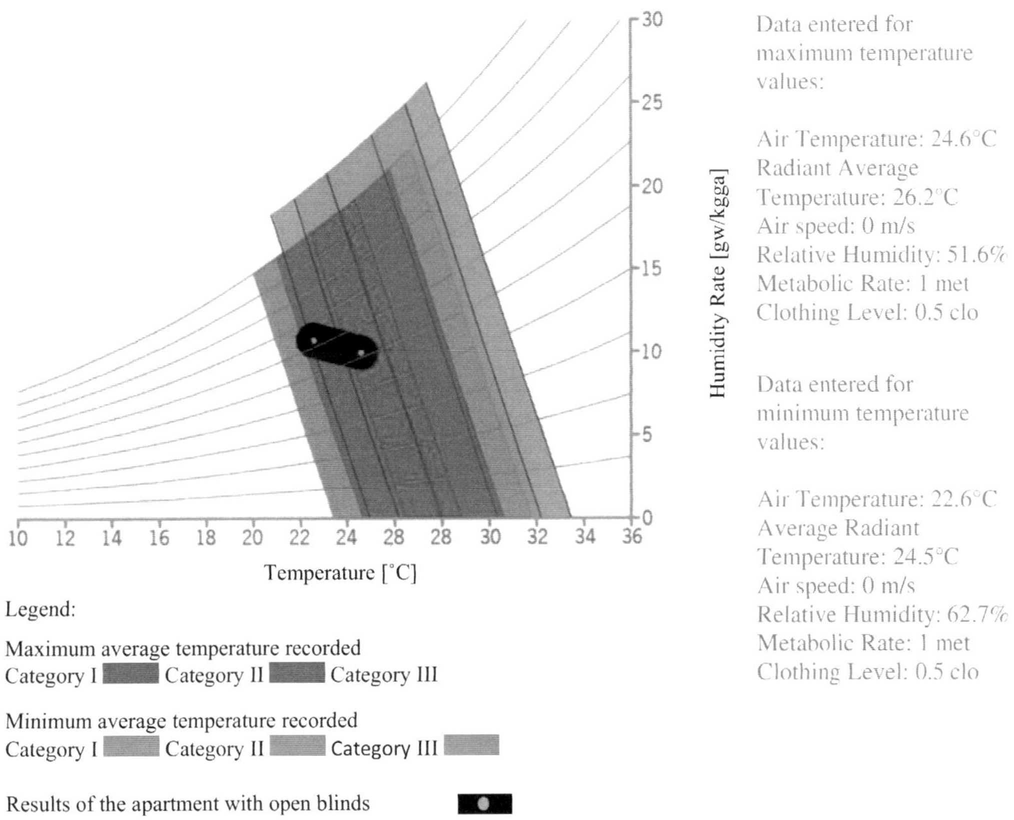

Figure 8. Evaluation of the comfort in the apartment with the blinds open, from 21 to 29 May.

4. Thermal comfort simulations

The collection of thermal comfort data in the building apartment allowed to conclude, as mentioned in the previous point, that even in conditions where the blinds are fully open for a week, the values comply with the standards. Due to the fact that the building is presently unoccupied, and that an occupation scenario would be difficult to monitor in a representative and controllable way, it was decided to simulate a situation of occupation of the apartment, establishing occupant's constant heat exchanges with space and electrical equipment that contribute to the heating of space as televisions and lamps i.e. the usual occupancy.

Simulation tools are important for testing something that is currently impossible to test since the building is unoccupied. DesignBuilder, a graphical interface available for the EnergyPlus (U.S. Department of Energy) energy simulation method. The program has BIM interoperability, facilitating the import of 3D models. DesignBuilder is a versatile computer program that provides several simplified options for the early stages of the project and more complex options that allow for a more detailed analysis of the energy model. The program allows the analysis of energy consumption, carbon dioxide emissions, comfort level of occupants, lighting, computational fluid dynamics (CFD), environmental impact assessment, among others.

The program uses EnergyPlus the U.S. DOE building energy simulation program for modelling building heating, cooling, lighting, ventilating, and other energy flows. It builds on the most popular features and capabilities of BLAST and DOE-2 but also includes many innovative simulation capabilities such as time steps of less than an hour, modular systems and plant integrated with heat balance-based zone simulation, multizone air flow, thermal comfort, and photovoltaic system. The simulations were carried out through the EnergyPlus program through the DesignBuilder, where the typological characteristics of the apartment type were inserted (the middle apartment or the one in which the floor and ceiling slabs have no thermal losses and are thus in the ideal conditions). Complementing the typological model were also inserted all the materials that are part of the construction system of the apartment, as well as the frames sized and characterized according to the existing one.

After inserting all the characteristics of the building, the operative temperature, ambient temperature and relative humidity were simulated, but to obtain these results, it was necessary to first find a day with the maximum and minimum temperatures, 25 and 15.5 ºC outside in order to match the measured "in situ" behavior of the building and thus validating the simulation conditions.

The objective of the simulated values in the program is to find the category of the apartment taking into account the adaptive comfort method preconized on the EN-16798 standard. The adaptive comfort method is only valid with some conditions such as the lack of mechanical refrigeration equipment, being priority given to the opening and closing of windows. Occupants should be on sedentary physical activities with metabolic rates ranging from 1.0 to 1.3 met. And there may still be variables of the inhabitants' clothing, that is, more clothing in winter, less in summer. This method is important because in order to create the basis for minimal and optimized intervention, the fact that people adapt to the existing one is relevant to maintain the characteristics of offering identity to the from the implementation to the materials used in the construction system

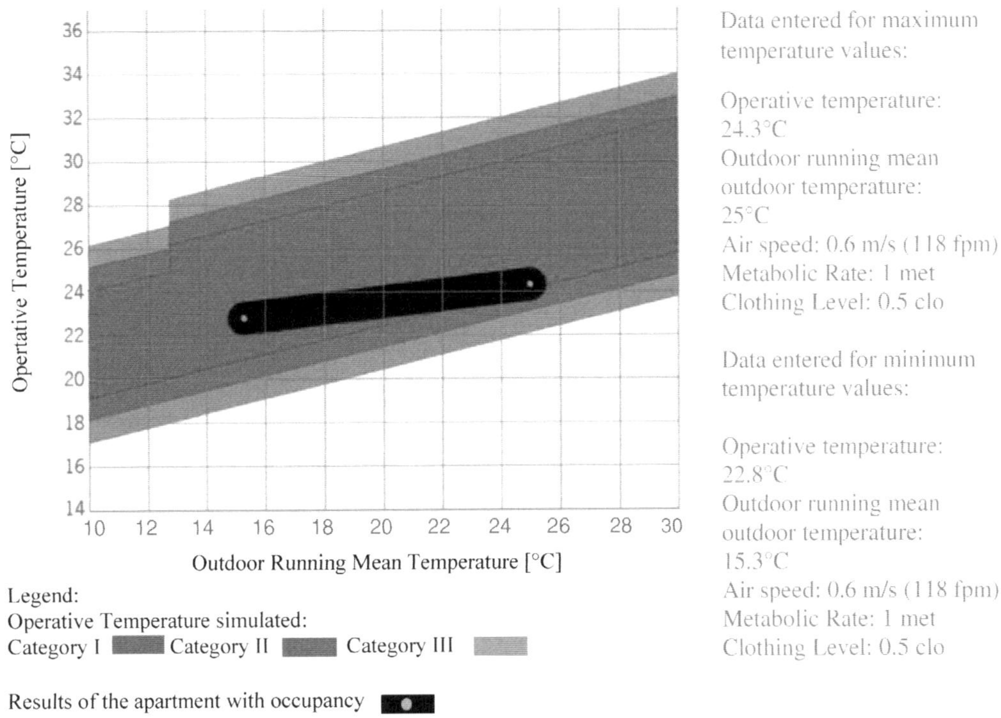

Figure 9. Evaluation of the simulated comfort in the apartment with occupation, from 21 to 29 May. EN-16798.

The interpretation of Figure 9 graph allows us to state this in an ideal scenario of occupation and taking into account the adaptive comfort requirements of the EN-16798 standard, with all the restrictions presented above. Edifício Lino is in Category I within the acceptable limits with Operative temperature from 20.8 to 25.8 °C i.e. only 5°C of temperature asymmetries.

5. Conclusions

"(...) the work of architecture is not only the demonstration of the artistic value of the architect. The architect will seek, as an artist, to create a work of art. But this work of art has to be useful and serve. Otherwise it will be sculpture, monument, but not architecture." [6]

Losa and Barbosa founders of ODAM (Organização dos Arquitectos Modernos), an organization of Oporto´s architects that at the 50´s defended the values and principles of Movement Modern, in particular the need to integrate comfort and functionality with the aesthetic and avant-garde values ofarchitecture. As such, constructive and material options, should be integrated on the proper design of the building. As referred by Maia, "at Edifício Lino, the architects were able to express their formal and spatial ideas, applying the common construction lexic in order to guarantee, passively, comfort of all the interior spaces [7].

The results obtained in the "in situ" measurements complemented by the simulations of the Edifício Lino proves the excellent performance of this kind of buildings – designed without any mechanical support, planed with passive ventilation strategies, that benefits of a good solar orientation and facades and window frames take advantage of and control solar incidence.

Despite the poor condition of the construction, abandoned and unused for decades, our study proves that with cleaning and simple reparation of the constructive elements as well as the implementation of a necessary maintenance we can prolongate the life of the building and match current comfort standards. The proof of this are the results obtained in the "in situ" measurements complemented by the simulations. The operating temperatures are, however, those that, through data collection and simulations, allow the building to be in category 1 of adaptive comfort where the difference in indoor temperatures should not exceed 5°C throughout the day. Mendonça, points that in the construction of buildings, "if there is no" strict application of principles, rules or standards that promote the rational use of energy and the conscientious introduction of new technologies, these levels of thermal comfort will tend to be achieved with greater use of mechanical heating and cooling systems, which will increase energy consumption in the sector, hence the pertinence of optimizing bioclimatic design." [8] An environmentally conscious design has the ability to respond to contemporary concerns and create the best conditions of comfort without mechanical supports or the implementation of new materials to optimize insulation.

Our measurements complemented by the simulations have proven that the capacity of the Edifício Lino, as it is can be inhabited according to nowadays expectations related to adaptive comfort in summer. To improve its comfort performance, the building only needs to be occupied, inhabited and used daily, perhaps this is the greatest achievement and what we must learn from the architects who designed these time challenging buildings.

6. References

[1] Tostões A 2009 *The importance of Docomomo in the intervention of the Contemporary Heritage*. (Seville Andalusian Institute of the Historical Patrimony) (Retrieved in http://www.youtube.com/watch?v=i8qOaVLv_Ms) on March 24, 2020
[2] Lima V 1973 The Portuguese Housing Problem in Barbosa C *ODAM: Organization of Modern Architects: Porto, 1947.1952* Porto Fundação Calouste Gulbenkian 25 (in portuguese)
[3] Vital L 1973 The House, Man, and Architecture in Barbosa C *ODAM: Organization of Modern Architects: Porto, 1947.1952* Porto Fundação Calouste Gulbenkian 37 (in portuguese)
[4] Tostões A 2015 Modern Heritage: conservation and reuse as a sustainable resource Revista Joelho nº6. (Coimbra EDARQ) 17 (Retrieved in: https://impactum-journals.uc.pt/joelho/article/view/_6_12/1779) on March 14, 2020
[5] Silva S 2009 Sustainability and Comfort of Buildings University of Minho School of Engineering. Guimarães 40 (in portuguese)
[6] Losa A 1951 From the Architect's Profession vol. XI, nº98 VÉRTICE Coimbra 538 (in portuguese)
[7] Maia, C 2018 The constructive idea. José Soares Apartment Rehabilitation Project of Edifício Vouga/Soares & Irmãos. Testing of methodologies and intervention processes on modern built heritage School of Architecture of the University of Minho Guimarães 210 (in portuguese)
[8] Mendonça, P 2005 Living under a second skin: strategies to reduce the environmental impact of passive solar constructions in temperate climates School of Architecture of the University of Minho Guimarães 12 Chapter II (in portuguese)

Acknowledgments

This work has support of Lab2PT - Landscapes, Heritage and Territory laboratory and FCTMEC and when applicable of the FEDER co-financing, in the aim of the new PT2020 partnership agreement.

IOP Conf. Series: Materials Science and Engineering 1054 (2021) 012011 doi:10.1088/1757-899X/1054/1/012011

Assessment of a Real-life Concrete Bridge Structure using Vibration-based Damage Detection Method

Bashir Ahmad Aasim[1, 2]*, Abdul Khaliq Karimi[1, 2] and Jun Tomiyama[1]

[1]Department of Civil Engineering & Architecture, Graduate school of Engineering and Science, University of the Ryukyus, Okinawa, Japan
[2]Civil Department, Faculty of Engineering, Kandahar University, Kandahar, Afghanistan

*bashir.aasim83@gmail.com

Abstract. Structural damage is usually presented in the form of decrease in stiffness resulting changes in the dynamic behavior of the structure, and natural frequency of the structure being one of the dynamic variables. A vibration-based damage detection method was employed to evaluate the structural health condition of a deteriorated bridge. The bridge has been closed for the traffic due to owing the extreme deteriorated health condition because of the immense existence of chloride ion in the airborne particles of the saline environment. This study presents, the comparison of natural frequencies of the year 2017 to that acquired in the year 2016 using vibration-based damage detection method. Even though the bridge was not exposed to the service load during the study period, yet, the results manifested decrement in natural frequencies revealing decline in structural stiffness in this period.

1. Introduction

Bridges are expected to have long service period among civil infrastructures. During this period, they are inevitable to suffer from different aspects such as Material aging, long term loading and environmental effects [1,2]. The recent catastrophic collops of the bridges confirmed lake in structural maintenance practices as it relates to bridges and, also highlighted the seriousness of the problem [3]. The importance of the bridge structures comes from the fact that they are one of the vital section of transportation infrastructure [4] which do not only affect economic, social and other various undertaking of a country but also have a huge an impact on safety of the people and their properties. Structural health monitoring systems have been widely utilized on engineering structures, especially bridge structures [5] Furthermore, corrosion of concrete reinforcement members has been acknowledged as one of the predominant deterioration mechanisms for steel reinforced concrete structures, whereas, reinforced concrete structures constitute the majority of civil infrastructures of the world. Therefore, SHM for concrete structure has attracted a lot of attention. A wide variety of techniques have been reported in the literature that may be suitably employed for non-destructive evaluation and detection of damage.

In this study, Vibration-based damage detection method is employed to evaluate structural health condition of a four-span reinforced concrete T-beam bridge located on a small river on the seaside in Nago city, Okinawa Prefecture, Japan, as shown in figure 1. The bridge was constructed and opened for the traffic 58 years back in the year 1962, and looking to be sound outwardly, has suffered severe deterioration underneath due to exposure to the aggressive environment in form of large salt mass accumulation at the bottom extent of the bridge girders. As mentioned, corrosion of embed steel is one of the main causes of deterioration. In this case, most of the reinforcement bars are exposed and corroded by chemical attack due to existence of large-scale cracks in concrete as seen in figure 2, and the main cause of deterioration is the existence of chloride ion in the immense saline environment. Taking its critical health condition into consideration, the bridge has been out of service in the past few years. However; evaluation results show the structure being still in the state of deterioration. Most of the time, environmental factors and lake in structural maintenance over time are considered to be the main causes of deterioration by mean of influencing load bearing capacity and durability of an engineering structure. The aggressive environment was able to deteriorate concrete and steel, since steel is very susceptible to the saline environment and the airborne particles were able to reach the reinforcement bars that resulted in corrosion of steel bars. This deterioration of concrete and corrosion of reinforcement bars could decrease load bearing capacity, bending stiffness and other factors of the bridges. For this purpose, VBDD assessment was conducted to illustrate the variation occurred in the structural dynamic behavior which usually leads to increase in flexibility and decrease in modal parameters notably in the natural frequency of a structure. Natural frequency is one of the widely used diagnostic tool for damage detection of a structure, which can be easily obtained and identified from modal tests with satisfactory level of accuracy [6].

Figure 1. Overall view of the bridge.

Figure 2. Deteriorated state of the bridge.

2. Experimental program

2.1. Outline of the bridge

The study was performed using one of the oldest T-beam concrete bridge in the area which was constructed on a small river near the seaside in 1962. The bridge consists of four equal spans of 9.25 m x 6 m with the total length of 37 m. The schematic of the bridge illustrating sectional dimensions is shown in figure 3.

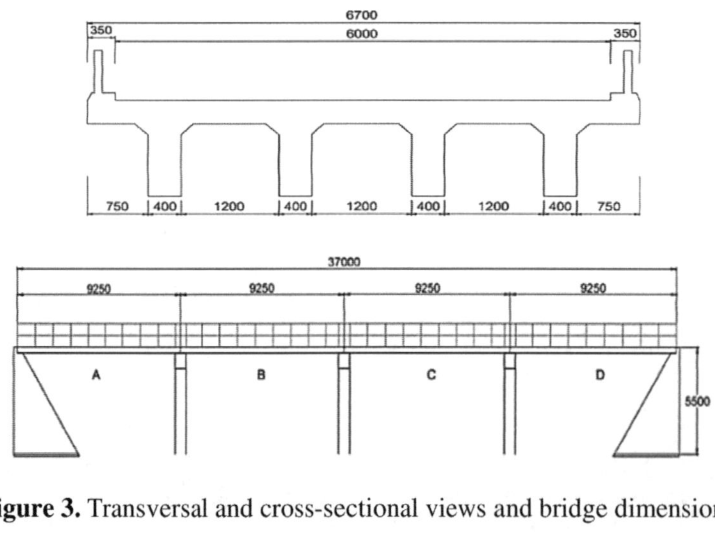

Figure 3. Transversal and cross-sectional views and bridge dimensions.

2.2 Experimental setup
To demonstrate the concept of an integrated damage detection system, vibration-based damage detection method is set up to evaluate the target bridge as shown in figure 4. The bridge is assessed several times during the period of few years. The aim of this evaluation is to compare the result of the year 2017 with that obtained in 2016. A drop-load excitation vibration method is performed to present the perturbation occurred in natural frequencies of the bridge in the period of one year.

2.3 Data acquisition
For measuring acceleration response, AR-10TF-type transducer of TOKYO SOKKI, which can measure acceleration in three directions up to 10g with a frequency range of DC-180Hz is placed on the top-mid surface of monitored span and is connected to a personal computer through EDX-11 (Kyowa) using DCS-100A software. The free vibration response (time domain waveform) is obtained from each span within the period of ten seconds by dropping a mass of 30kg from 70cm height on the middle of the span next to the accelerometer shown in figure 5. These time domain waveforms are transformed into frequency domain waveforms by Fast Fourier Transform (FFT) using DAS-200A data analyzing software. FFT is the fast-computational algorithm for Discrete Fourier Transfer (DFT) and the basic concept underlying the FFT has to do with taking an array of time-domain waveform samples and processing them to produce a new array of frequency spectrum domain samples. Frequency peaks in the resulting frequency spectrum are used to evaluate the integrity of structural components because the integrity of structural members affects their cross-sectional vibration modes and the corresponding natural frequencies [7].

2.4 Evaluation procedure
To record the vertical acceleration response of the bridge, the assessment procedure starts from span A to span D accordingly as shown in figure 6. The sensor is placed in the center of each span and a sandbag of 30 kg mass is hanged and released perpendicularly on the span to generate free vibration waves.

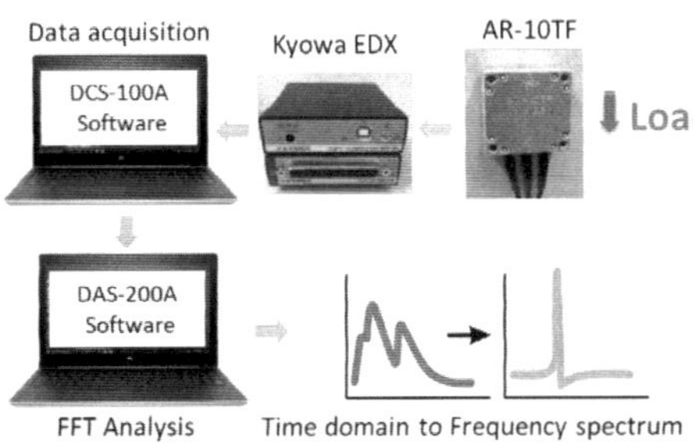

Figure 4. Data acquisition system.

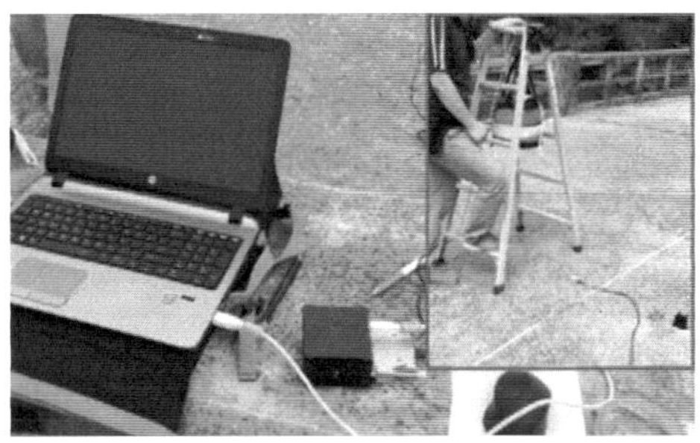

Figure 5. Method setup and bridge assessment.

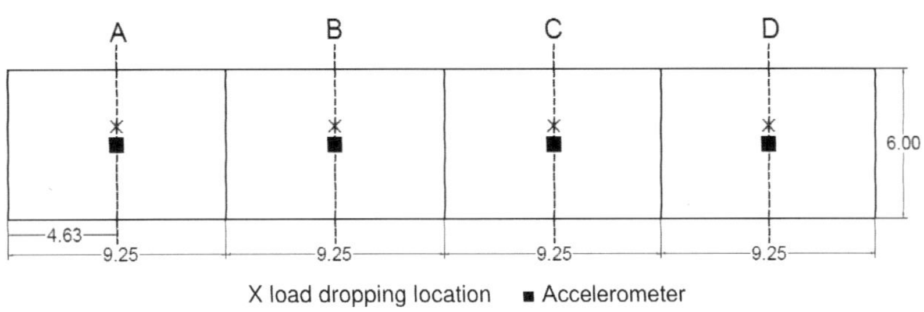

Figure 6. Accelerometer and load dropping positions at the time of bridge evaluation.

3. Results

3.1 Results of the year 2016

The evaluation was carried out in 2016 for acquiring the vibration from each span separately A, B, C and D respectively and, the results are presented in the form of time domain and frequency spectrum graphs as shown in figure 7. Legends X, Y and Z present acceleration in three directions with Z being vertical component of acceleration which illustrates the health condition of the structure. Time domain waveforms and natural frequencies of the assessed spans are presented illustrating the health condition of the referred spans. The frequency peaks are used to demonstrate the integrity of the structural components.

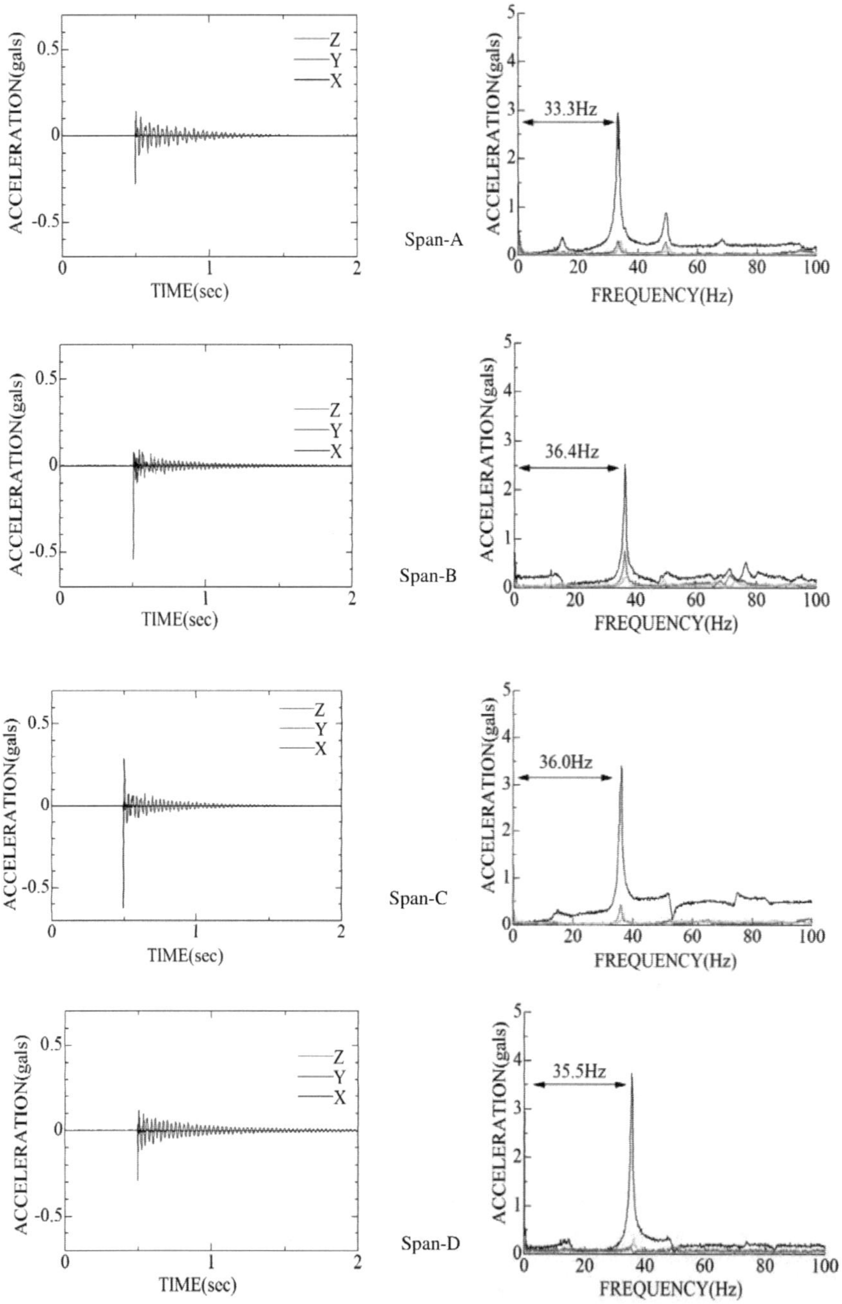

Figure 7. Time history and frequency domain of the bridge.

3.2 Results of the year 2017

Graphs in figure 8 show the vibration response of each span after a period of one year in 2017. The data obtained from the bridge was compared with that acquired in 2016. Decrease in frequencies confirmed decrease in serviceability of structural members of the bridge.

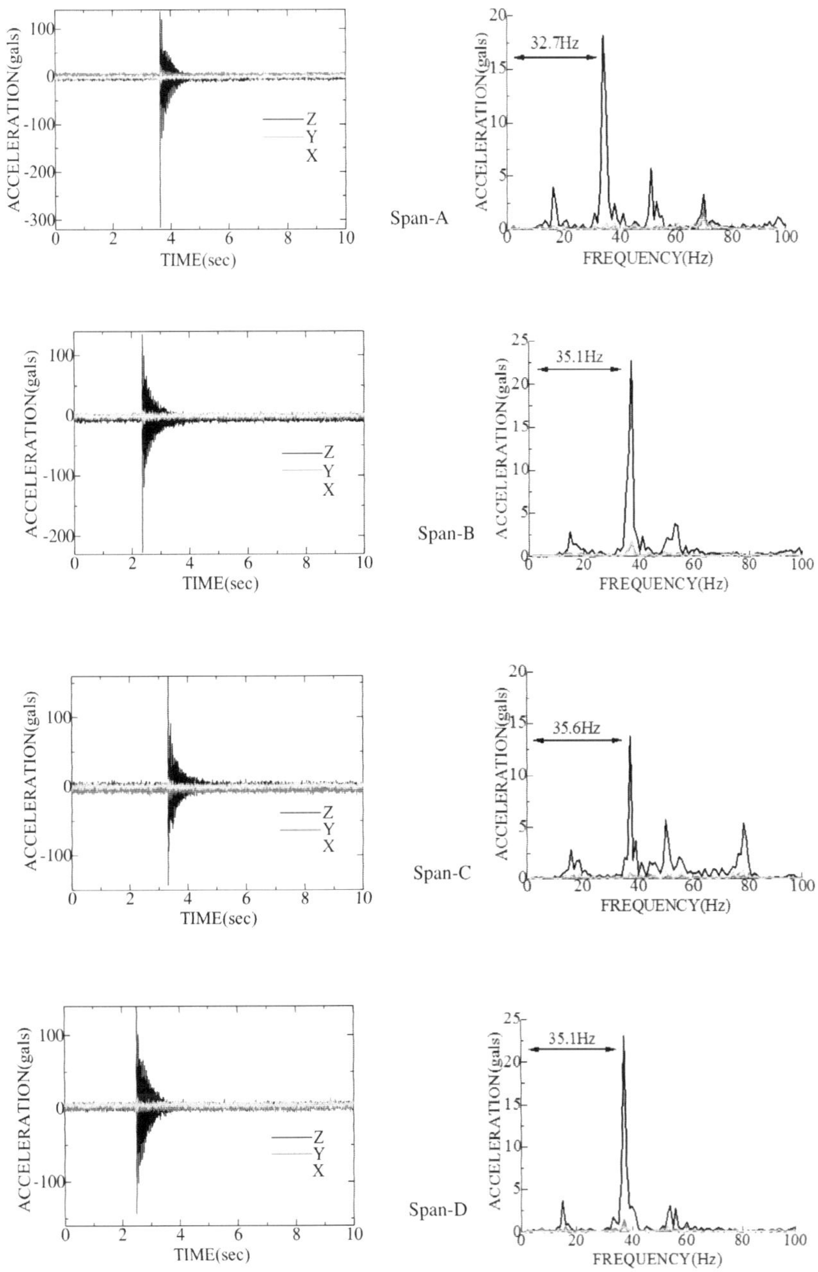

Figure 8. Time history and frequency domain of the bridge.

4. Discussion

The overall results showed degradation in natural frequencies leading to decrement in stiffness of the structure corresponding to the deterioration experienced by the structure throughout the period of one year. The reason of increase in flexibility is the existence of cracks which gradually lengthened and widened till the airborne particles reached reinforcement bars. Each span of the bridge is evaluated in the year 2016 and compared with that for 2017 to study variations in natural frequencies. The natural frequency of the bridge structure at spans A, B, C and D changed from 33.3Hz, 35.5Hz, 36Hz and 36.4Hz in year 2016 to 32.7Hz, 35.1Hz, 35.1Hz and 35.6Hz in 2017 respectively, as shown in figure 9. Among all the spans, the highest decrement observed at span C (0.9 Hz). The average natural frequency of the structure obtained in the year 2016 to be 35.3Hz, which reduced to 34.625Hz in 2017 presenting the descendant trend of the natural frequency of the bridge structure is shown in figure 10.

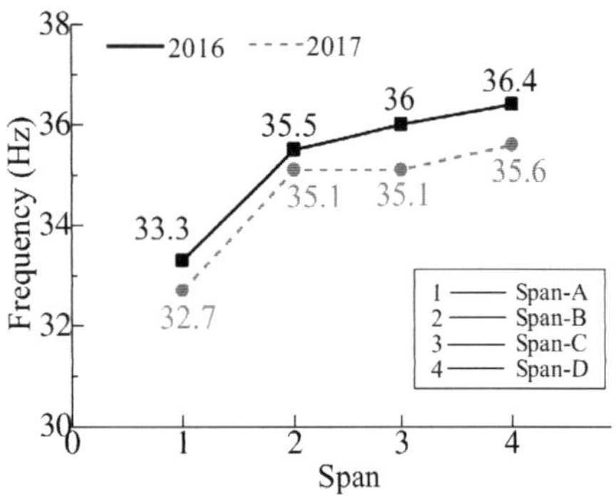

Figure 9. Stiffness decline in each span.

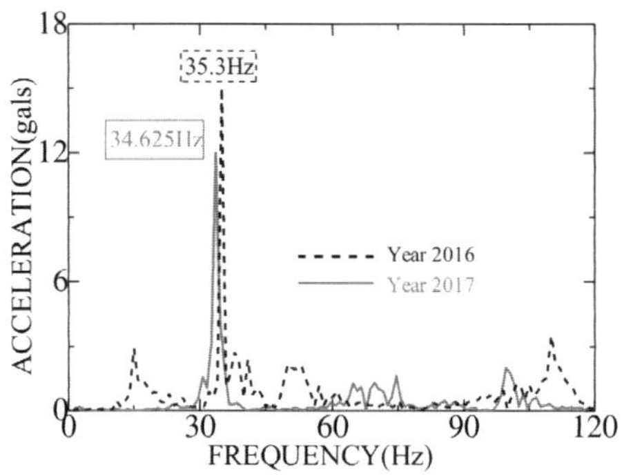

Figure 10. Overall reduction in stiffness of the bridge.

5. Conclusion

Vibration characteristics of a reinforced concrete bridge were monitored using potable vibration measurement system with a load drop impact excitation. The changes in natural frequencies of the bridge structure during the period of one years were adopted to illustrate the health condition of the bridge. The bridge even being out of service during the study period manifested decrement in structural stiffness in the form of changes in natural frequencies detected by the vibration based non-destructive testing method, rooting from severely saline environment. The proposed method was found to be sufficient as well as convenient for structural damage detection in terms of using dynamic characteristics as a diagnostic tool.

6. References

[1] Li, Hui. (2004). Recent Advances of Structural Health Monitoring in Mainland China.

[2] Karimi, A. K., Aasim, B. A., Tomiyama, J., Suda, Y., Aydan, Ö., & Kaneda, K. (2020). Experimental and numerical studies on the control of horizontal cracking at the ends of hollow-type pretensioned girders. *SN Applied Sciences*, *2*(10), 1-17.

[3] M. Moore, B. Phares, B. Graybeal, D. Rolander, G. Washer, Reliability of visual inspection for highway bridges. U.S department of transportation Federal Highway Administration 2001.

[4] Karimi, A. K., Jaheed, A. B., Aasim, B. A., & Farooqi, J. A. (2019). Structural Condition and Deficiencies of Present Constructed Bridges over Zahirshahi Canal and Proposal of New Design Using AASHTO Codes. World Journal of Engineering and Technology, 7(02), 325.

[5] Aasim, B.A., Karimi, A.K., Tomiyama, J. et al. Numerical verification of accelerometer-based assessment of hollow-type pretensioned concrete girder. Asian J Civ Eng 21, 437–447 (2020).

[6] Y. Luo, Z. Wang, B. Xu, Experimental damage identification in concrete structure using stack migration imaging technology, J. Adv. Conc. Tech. 10 (2012) 41-46.

[7] B. Aasim, A. Karimi, J. Tomiyama, Ö. Aydan, Y. Suda, Detection of damage in concrete structure via shifts in natural frequency, Int. J. Tech. Research. Applica. 5 (2017) 48-52.

Visual Inspection and Determining Bridge Load Rating over the Torrential Rimac River in Lima, Peru

C Blas[1], F Fernandez[1] *and E Carrera[1]

[1] Faculty of Engineering, Universidad Peruana de Ciencias Aplicadas, Prolongación Primavera, 2390, Lima, 15023

*Corresponding author: u201510159@upc.edu.pe

Abstract. Nowadays, the appropriate functioning of road infrastructure is a major concern that governments face. One of the structures most vulnerable to failure are bridges. In Peru, the hydrodynamics of torrential rivers, attributable to its geography, generates the collapse in the bridges by scour. Likewise, the increase in service overloads, compared to design loads (HS20), united with the fact that many bridges have more than 50 years in service, is another common factor that contributes to the bridge failures and should be considered in the evaluation of bridges. Therefore, this research aims to inspect, evaluate the load capacity in the superstructure of the Dueñas bridge using the LRFR methodology and propose improvements in the maintenance process. This bridge is made of prestressed concrete built in the 60's and over the torrential Rimac river in Lima, Peru. The findings, according to the field visit, indicate the main problems are scour in the intermediate pier and damage in the concrete in the structural elements. Additionally, the rating factor (RF) calculated in the beams do not exceed to unit which implies that they only need structural reinforcement. Finally, improvements were proposed for the prevention of failure in the beams and scour in the substructure.

1. Introduction

Road infrastructure assume an important role in countries, since it promotes social integration and sustained economy growth in modern cities. The successful performance of this transport infrastructure is still a major challenge for responsible governments around the world. One of the most important structures are the bridges, due to their purpose of connecting roads achieving the objective of road network. Bridges are highly vulnerable to failure caused by various factors, such as aggressive environmental effects and overloads, which have an impact on structural and functional deficiency, safety and integrity of the population [1].

Peru is a country that, for its geographical location, presents several natural phenomena which cause severe damage to bridges. The El Niño phenomenon is the most frequent cause of bridge failures in relation to rivers behavior: the scour. This consists of removal of materials or solid waste from the riverbed, which the main agent is water in the surroundings of the bridge substructure. It arises when appears local speeds greater than average speed of current produced by vortices in different directions in the substructure and is originated for two main reasons [2]. On the one hand, Peruvian geography facilitates the formation of torrential rivers through steep and narrow valleys at high altitudes, causing changes in flow velocity and channel width [3]. On the other hand, the El Niño phenomenon represents a violent and unusual change in climate, allowing an increase in the discharges of rivers and streams due to the high intensity of rains with long durations between December and March, in addition to the lack of "Prevention culture" and insufficient river drainage [4]. For example, environmental conditions caused by the El Niño phenomenon in 2017 not only generated 1.9 million Peruvians affected, but also major losses and damage to the infrastructure of 601 bridges affected by scouring and would require USD 15 million to repair them or build them to 168 bridges [5].

Furthermore, other common factor that contributes to bridge failures is the union between the increase in service overloads, with respect to design loads (HS20) and that the bridges have more than 50 years in service. In urban areas with high population density such as Lima, structures are constantly affected by this combination of factors affecting their long-term capacity. For instance, in relation to the increase in the volume of the vehicle fleet, Ministry of Transport and Communications (MTC) shows that heavy and light vehicles increased by 12% between the years 2017-2018[6].

Therefore, this research aimed to inspect the actual condition of the elements and structurally evaluate the superstructure of the Dueñas Bridge over the Rimac river in Lima, Peru, as well as propose improvements to prevent failure in the beams and scour in the intermediate pier. In addition, this bridge has not suffered damage by seism during its service life; thus, the study does not consider seismic effect for this evaluation.

2. Materials and method

2.1. General description of Dueñas bridge

The bridge under study, as shown in figure 1, is located in the urban area of Lima, Peru and was built in the 60's. It is made of prestressed concrete with a total length of 70 m, which is divided into two simply supported sections of 35 m supported by two abutments and a central pier. The height from the bottom edge of the footing to the top edge of the deck is 18 m.

Figure 1. General view of Dueñas bridge.

2.2. Inspections

MTC has a manual for inspection through standardized activities and procedures to determine the current physical and functional condition of the components in order to prevent deterioration. Generally, inspection is a task that involves two processes: visual and physical. In addition, they are classified three inspection types (initial, routine and special), which the first two are necessary for vehicle load capacity evaluation and, therefore, they will be used in the case study.

2.2.1. Initial Inspection. It's the first inspection that allows the collection of information necessary for a database when the bridge is not registered in an inventory. It consists of collecting information through plans, files, history of the bridge, previous inspections or information regarding the bridge. This information will be recorded in an inventory by structure with relevant details found. MTC considers as a minimum to include the following information in the database: general data, basic information, geometric design characteristics, previous inspections, rehabilitation history and photographs.

2.2.2. Routine Inspection. This inspection is known as field inspection and carried out to detect timely deterioration, locate and repair them to prevent an increase in damage severity. This part implements the use of datasheets that include photographs of the damage, recommendations report for maintenance or if a special inspection is necessary. In particular, greater consideration should be taken of damage involved as: impact on user safety and functional or resistant characteristics degradation in the bridge components [7].

2.3. *Load and Resistance Factor Rating evaluation of the superstructure*

Primary focus of this section is the assessment of the safety of bridge for live loads (including overload) and fatigue. The LRFR method, provided by the Manual for Bridge Evaluation (MBE) in Chapter 6, is one of the last methods used based on the Load Resistance Factor Design (LRFD) philosophy. It is mainly used to determinate the safe live load carrying capacity in a newly designed or existing bridge structure. The factor (RF) is calculated using LRFR method from information obtained of bridge plans and supplemented with information collected from field inspections. The RF decides which structures have substandard load capacities that may require posting or other remedial action. The LRFR load rating equation is shown below [5]:

$$RF = \frac{C-(\gamma_{DC})(DC)-(\gamma_{DW})(DW)}{(\gamma_{LL})(LL+IM)} \tag{1}$$

where, for limit states of resistance,

$$C = \phi_C \phi_S \phi R_n \tag{2}$$

$$\phi_c \phi_s \geq 0.85 \tag{3}$$

The RF denotes the rating factor. C is the member's capacity or factored resistance. Rn represents the nominal strength of the member in the LRFD code and is calculated from the inspection condition. DC, DW and LL+IM denote the effects of load due to the weight of structural components and accessories, the weight of the wear surface and utilities, other permanent loads, live load plus dynamic allocation, respectively. γ_{DC}, γ_{DW} and γ_{LL} are the corresponding load factors. φ_c, φ_s and φ are the condition factor, system factor, and strength factor respectively [8-9].

3. Results and Discussion

3.1. *Initial inspection*

The cross section has a total width of 20.6 m subdivided into 4 two-way lanes of 3.6 m in length each. The figure 2 shows the total cross section in which the six beams were named with a number. The figure 3 is represented as a semi-cross section where the thickness of the asphalt layer is 5 cm, sidewalk is 2.5 m wide and the steel railing is 1 m high. Regarding the beams, they are made of prestressed T-beam bridge decks whose dimensions are variable section whose maximum value is 0.40 m and 1.70 m in height.

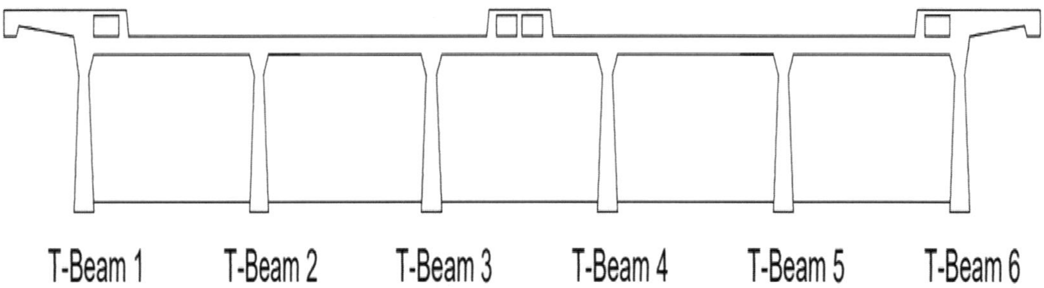

Figure 2. Cross section of Dueñas bridge.

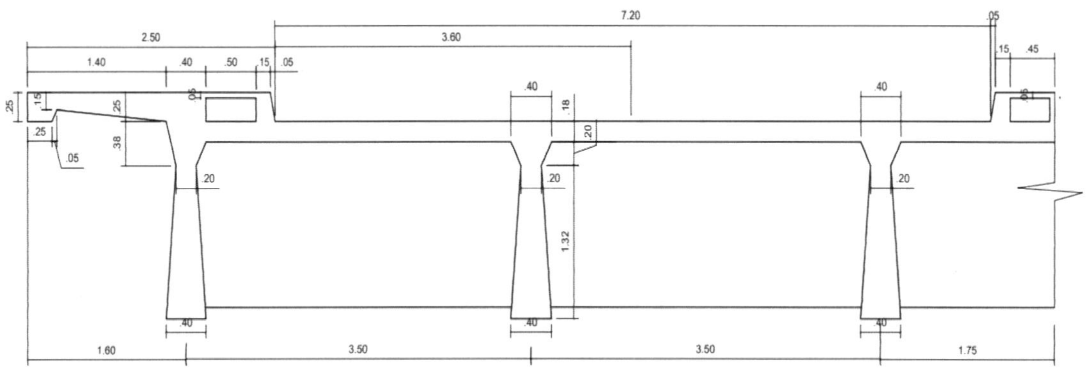

Figure 3. Semi-cross section of Dueñas bridge.

3.2. Routine inspection

For this step, two maps of different views were developed in order to locate the damage to the structure during field inspection.

The first map, shown in figure 4, has five main problems. In the substructure, scour was found as the most prominent damage and advancing in the central pier. The actual scour height with respect to the level of the ground in 1965 is 2.63m (number 1). However, although there is no problem of scour in the abutments, this element is surrounded by garbage filling and toxic waste (number 4). The water of the Rimac river is contaminated with sewage from the invasion of houses on the river bank, mining and construction waste. In addition, the starling section is eroded (number 3) and vertical cracks were evident in the lower area of the pier (number 2). Finally, the T-beams and deck presents cracks in both directions, exposure of the steel in one of them as a result of the minimal coating used, and efflorescence due to high humidity (number 5).

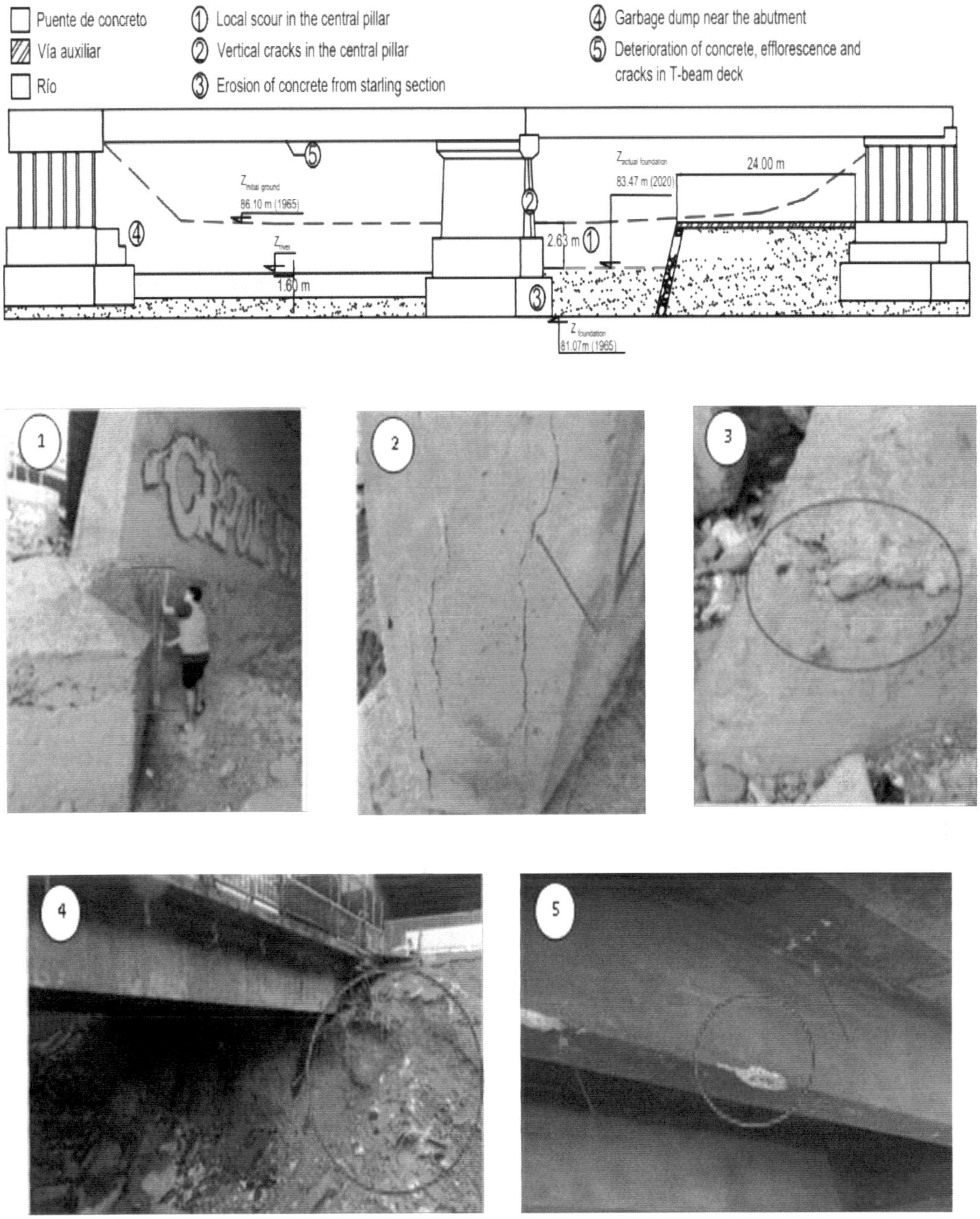

Figure 4. Damage map for the overview of Dueñas bridge.

The second damage map was made for the plan view which reflects four problems related to the users' safety according to figure 5. For example, numbers 2 and 3 affect the pedestrian because there are cracks in the sidewalks on both sides and rust is evident on the railings. For numbers 1 and 2, it not only affects drivers and vehicles, but also deteriorates the structure, since the pathology called "crocodile skin" and the poor condition of the expansion joints can alter the bases of the pavement.

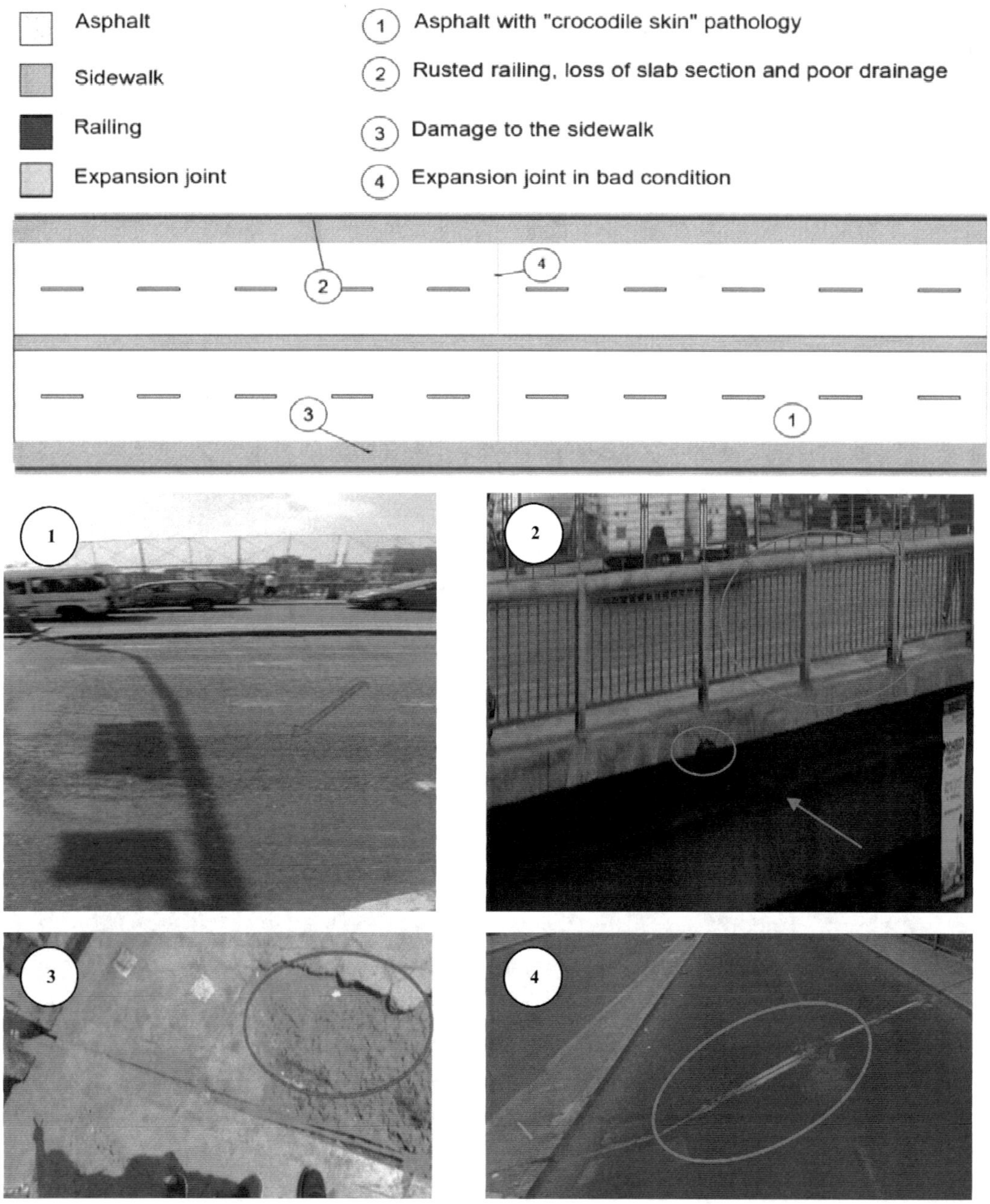

Figure 5. Damage map for the overview of Dueñas Bridge.

3.3. *Rating factor (RF)*

According to the MBE, they note that, if the RF at inventory level is less than 1.00, it needs to be evaluated at the operation level. Additionally, in Peru, bridge evaluation experts recommend that, if the RF value is greater than 1.00, the bridge is optimally state. If it varies between 0.65 and 1.00, you will need structural reinforcement and, if the RF value is less than 0.65, the bridge must be demolished.

The structural evaluation, based on the LRFR methodology, shows that the exterior T-beams are more critical than the interior T-beams because the external T-beams (1 and 6) support the pedestrian live load, in addition of the dead load distribution. This pedestrian live load is high due to the geometry of the sidewalk which has a width of 2.5 m. In table 1, all the parameters used to calculate the RF of the internal and external T-beams are presented according to equation (1) for the load combination Resistance I. Additionally, the RF is calculated for the inventory level and the operation level. For the inventory level, table 1 shows that none of the T-beams have enough capacity to withstand bending stresses. The flexural RF of the external T-beams are less than 1.00 in 30%, while the average of the internal T-beams is less in 23.5%. For the operation level, table 1 shows that no T-beam has enough capacity to withstand the bending stresses either, since on average the external T-beams are less than 1.00 by 9% and 1% for the internal T-beams. Therefore, all T-beams only require structural reinforcement.

Table 1. Parameters and RF values for the Dueñas bridge.

T-Beam	1 and 6	2 and 5	3 and 4
RF(Inventory Level)	0.71	0.77	0.76
RF(Operation Level)	0.92	0.99	0.98
DC (kN·m)	3364	4001	4129
DW (kN·m)	49	520	451
$LL+IM$ (kN·m)	5139	3923	3923
R_n (kN·m)	11219	11631	11631
ϕ	1.00	1.00	1.00
ϕ_c	0.95	0.95	0.95
ϕ_s	1.00	1.00	1.00
γ_{DC}	1.25	1.25	1.25
γ_{DW}	1.50	1.50	1.50
γ_{LL}	1.75	1.75	1.75
γ_{LL} (Operation Level)	1.35	1.35	1.35

4. Improvement proposal

The proposals to improve the Dueñas bridge have been divided into two parts. The first part is proposed as a result of the routine inspection developed in section 3.2. Thus, as general measures of prevention of bridge failure, it is recommended the following:

- To avoid the reduction of the width of the river through an adequate urban planning, since there are housing constructions and litter on the banks of the river in order to prevent the increase of the average flow velocity (local scour) presented in the foundations.
- Elimination of the existing road under the bridge so as not to reduce the width of the river.

- Protection of concrete against acids and sulfates in river water.
- Application of fissure seals on concrete structural elements of the bridge.
- Construction of river defenses.

The second part is linked to the calculation results of superstructure structural evaluation carried out in section 3.3. Carbon fiber (Sika Carbodur S1012) is suggested to be applied to the beams (due to their low resistance capacity to current admissible loads) to improve the RF parameter. In figure 6, the location of the carbon fiber strips for the flex and shear reinforcement was rendered in 3D in Revit.

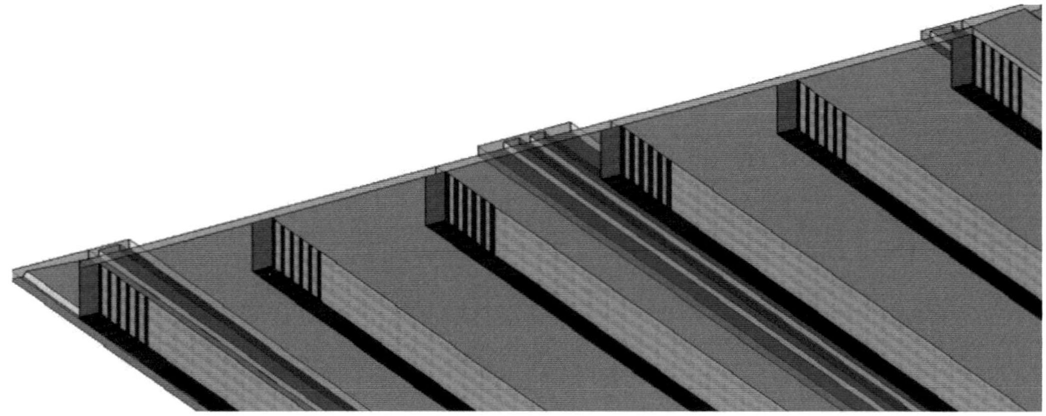

Figure 6. 3D view in Revit of the application of the carbon fibers bands in the beams.

For this purpose, in figure 7 (a), the cross-sectional view of the beam was represented in which the bands are placed on the center of the beams in all the bases whose dimensions are 0.30 m wide and a length of 30 m for the flexural reinforcement. Regarding to the longitudinal view, the placement of 6 bands of 0.10 m wide in the shape of "U" will be made every 0.20 m covered on each T-beam to ensure shear reinforcement shown in figure 7 (b).

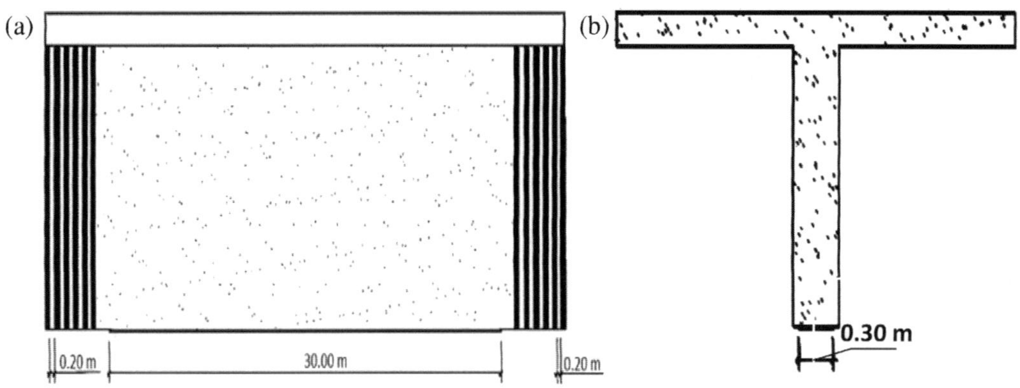

Figure 7. (a) View of the cross section of the application of the fiber at the base of the beam. (b) Longitudinal view of the use of carbon fibers in the beam.

Finally, a flow chart was developed, as shown in figure 8, which summarizes all the procedures executed in this investigation. It starts from basic inspections (gathering general information), routine inspection (in field and office) and the improvement proposals.

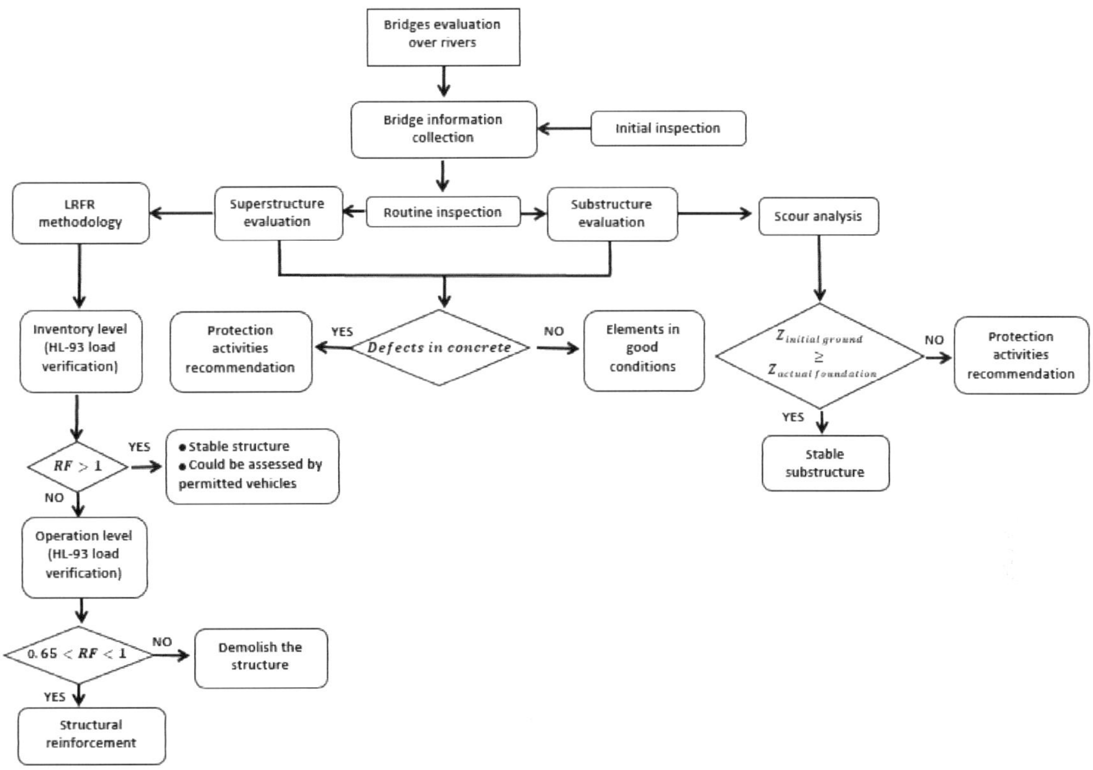

Figure 8. Flowchart for the evaluation of bridges over rivers.

5. Conclusions

In the present study, the visual inspection and superstructure evaluation, using LRFR method to the Dueñas bridge, was carried out with the objective of knowing the current physical and structural condition. It can be concluded that the Dueñas bridge presents grave problems in its current condition and does not have an inventory. Next, the most important findings and the improvement proposal to prevent the failure will be known:

- Before the visual inspection, it was necessary to collect information about the characteristics of the Dueñas bridge because the MTC has an inadequate database for bridges (a common characteristic in developing countries). The inventory has a fundamental role because it allows us to obtain information about any bridge. Also, it generates a broad vision on the evaluation and knowledge of previous inspections. Therefore, it is essential to have an updated inventory of the bridges in the construction and service life stage.
- The results of the evaluation using LRFR method indicate that the Dueñas bridge needs a structural reinforcement for all beams, since all RF values do not exceed 1.00. For this reason, it is recommended to apply carbon fiber to reinforce bending stresses and ensure shear force.
- During the routine inspection, damages were found associated to bridge and user safety. With regard to the bridge, the main problems are scour and damages in concrete caused by behavior of the torrential river, while, for users, there are defects in sidewalks, asphalt, and rusty metal railings. Consequently, it is important that the authorities have proper and opportune maintenance management to prevent bridge failures.

6. References

[1] Frangopol D M, Dong Y, and Sabatino S 2017 *Struct. and Infrastruct. E.* **13** 1239-57

[2] Rocha A 2013 *Introduction to the Hydraulics of road works* (Lima: Institute of the Chamber and Management) pp 249-250

[3] Rocha A 1998 *River Hydraulics* (Lima: National University of Engineering) pp 12-17

[4] Rocha A 2011 *The impact of the El Niño phenomenon in urban areas*. Research Report (Lima: College of Engineers of Peru) pp 29-37

[5] America Economica 2017 *Reconstruction of bridges and roads in Peru, after a climatic phenomenon, would reach the US$3.731M* Retrieved on August 27, 2020 from https://www.americaeconomia.com/economia-mercados/finanzas/reconstruccion-de-puentes-y-carreteras-en-peru-tras-fenomeno-climatico

[6] Statistics Office of the Ministry of Transport and Communications 2018 *Statistical Yearbook 2018* Technical Report (Lima: Ministry of Transport and Communications) pp 142-143

[7] Ministry of Transport and Communications 2019 *Bridge Inspection Guide* (Lima: Peruvian Standard) pp 37 - 58

[8] American Association of State Highway and Transportation Officials 2011 *The Manual For Bridge Evaluation* (Washington: American Standard) pp 168 - 171

[9] American Association of State Highway and Transportation Officials 2014 *AASHTO LRFD Bridge Design Specifications* (Washington: American Standard) pp 28-118

IOP Conf. Series: Materials Science and Engineering 1054 (2021) 012013 doi:10.1088/1757-899X/1054/1/012013

Sustainable Urban Pavement for *Cities Affected by El Niño* using Porous Concrete

B Aguirre [1*], M Anchiraico [1], J Rodríguez[2], F García [3]

1 Bachelor, Civil Engineering Program, Universidad Peruana de Ciencias Aplicadas, Lima, Perú
2 Full Professor, Civil Engineering Program, Universidad Peruana de Ciencias Aplicadas, Lima, Perú
3 Professor, Civil Engineering Program, Universidad Peruana de Ciencias Aplicadas, Lima, Perú

*u201517072@upc.edu.pe

Abstract. The El Niño phenomenon is caused by the change in atmospheric pressures, which produce the accumulation of hot surface waters on the eastern flank of the Pacific Ocean; causing intense rainfall that runs over the surface affecting the urban drainage of the city due to the lack of a permeable pavement; porous concrete allows infiltration of surface water runoff through its pores. The present investigation evaluates porous concrete in the range of w/c relationships of 0.30 and 0.32; the results indicate that the compressive strength, flexural strength and permeability coefficient increase; and that the surface runoff, cost, water footprint and carbon footprint are lower than conventional concrete.

1. Introduction

Many countries have witnessed different catastrophes over the years, which have caused economic, social and environmental damage to populations, due to anthropic and/or natural changes caused by cyclones, hurricanes, droughts, and floods; the latter two being related to the La Niña and El Niño phenomena. El Niño is the product of a shock of sea surface temperature and air pressure that occurs in the equatorial Pacific Ocean [1], its frequency and intensity is related to climate change and the greenhouse effect, it manifests itself in a non-periodic [2] and has a duration of 7 to 16 months [3]; chronologically, it begins in 1957 and ends in 1958, then continues in the periods of the years 1965-1966, 1972-1973, 1982-1983, 1986-1987, 1987-1988, 1991-1992, 1997-1998, 2002- 2003, 2009-2010 and 2016-2017 [4]. Historically, the 1997-1998 event was the one that caused major economic damage, ranging from US $ 32 billion to US $ 96 billion in the countries of the Pacific Basin and the Tropics [5], and produced the greatest impacts negative effects on the sanitation of cities, causing damage to their water supply systems, sewage and rainwater [5], [6], [7].

These damages were produced by surface runoff, product of excessive rains, present in the impervious carriageways of the roads which were conceived without an integral technical conception of the engineering professionals between urban planning and drainage systems, and lack of resources [8]; generating damage to drinking water supply and sewerage systems, and an increase in surface temperature and floods [6]; given this situation, the possibility arises of using sustainable urban drainage systems (SUDS) to control floods and pollution generated in an integrated and sustainable way [9], being able to design a porous concrete (PC) permeable pavement to reduce and drain surface runoff from urban areas, as well as the effect of urban heat [10], [11]. This concrete that is placed in the wearing course, contains little or no amount of fine aggregate [12], presents a significantly greater amount of pores than conventional concrete (CC), whose void content is from 11% to 35% [13], its w/c ratio varies between 0.28 and 0.40 [14], it has a resistance to compression between 28.5 to 285.5 Kg/cm^2 [15], and its permeability coefficient is between 0.4 to 1.7 cm/s [12]. As drawbacks of the PC, we can mention that it is not applied to heavy traffic because its high content of voids does not allow it to reach the resistance required to meet the load requirements of high traffic [16]; it also presents the risk that, applied to permeable pavement, the infrastructure will fill up and degrade when an adequate maintenance program has not been followed [17].

One way to visualize the efficiency of this permeable pavement is by using computer programs developed for rain and runoff, which allow knowing the dynamic responses of the elements that make up an urban drainage system for a rain event [18]; among the tools used for SUDS we have: Model for Urban Stormwater Improvement Conceptualization (MUSIC) [19], Benefits of SUDS Tool (BEST) [20], System for Urban Stormwater Treatment & Analysis Integration (SUSTAIN) [21] and Stormwater Management Model (SWMM 5.1) [22], this being used in research because it incorporates a simulation module for sustainable drainage techniques [10].

Some investigations carried out on the mechanical behavior of PC for the w/c relation with respect to compressive strength, flexure and permeability coefficient at 28 days indicate that the compressive strength increases [16], being able to reach up to 12% for w/c of 0.30-0.34 [23], [24]; that the flexural strength increases [25], reaching a value of 12% for w/c between 0.30-0.34 [26], [25]; and that the permeability coefficient should not have an w/c less than 0.30 [24], the w/c of 0.32 being optimal to obtain a permeability coefficient greater than 2 mm/s [27]; [28] indicates that the w/c influences the permeability coefficient. In the economic field, we can indicate that PC can reduce the general costs of a project [29], when its manufacture is largely manual and can be done without heavy equipment, the cost is lower [30], and it is advisable to analyze the material costs [26].

In the environmental aspect, there are some studies on the sustainability of urban pavements, [31] indicating that the water footprint increases, due to the runoff of rainwater that flows through the paved surfaces; the excess of this water generates a lower value of the water resource, which causes water losses due to lack of control [32], recommending good management [33]. Regarding the carbon footprint, we can indicate what was expressed in [34], [35] who argue that the use of a permeable pavement is a good alternative to reduce greenhouse gas emissions, because by mitigating the heat island urban carbon dioxide (CO2) is reduced.

The city of Tumbes located to the East with respect to the Pacific Ocean and at 6 m.a.s.l presents a recurrence of hydrometeorological dangers, marked by the occurrence of floods due to the occasional presence of torrential rains [36]. The last rainfall was in 2017, whose intensity was 16 times more than the annual average of precipitations, it caused intense rains and floods that left considerable damage to: public institutions, health, education and homes [37]. A large number of homes were affected because they were located in the channels of streams and rivers, they were located in areas with a deficient storm drainage system, and because they were built without a drainage system for rain, generating destruction in the water pipes and drain [36]. To date, the city does not have an appropriate storm drainage system for the volume of precipitation of an event such as the El Niño phenomenon [38].

This research addresses the problem of the lack of a sustainable permeable urban pavement in cities affected by the El Niño phenomenon. For this, the compressive strength, flexural strength and permeability coefficient of PC were studied; and the unit cost, water footprint, carbon footprint, and storm runoff comparison between a CC pavement and a PC pavement using SWMM 5.1 software.

2. Materials and methods

2.1. Materials

Type V Portland Cement was used [39]; the aggregates were natural with a maximum size of Ø 3/4" [40]; ZRR PLAST-971 [41] Superplasticizer was used; the water was drinkable [42]. Surface runoff was simulated with an HP OMEN laptop with Intel (R) Core (TM) processor it-8750H CPU @ 2.20GHZ and SWMM 5.1 software [43]; it was also used for footprints [44].

2.2. Methods

2.2.1 Tests performed. The PC of f´c = 210 Kg/cm^2 was designed [39, 40, 41, 42], with w/c ratios of 0.30 and 0.32, coarse aggregate (95%) and fine aggregate (5%). Table 1 shows the mixes, preparation and curing for 28 days [45]; 4 specimens were used per test: compression (6 "x12") [46], permeability coefficient (4 "x8") [47], and flexion (6 "x6" x15 ") [48].

2.2.2. Unit prices. The data referenced in [49], [50] and [51] were used in the elaboration of the analysis of unit prices of the CC and PC.

Water footprint: The Declaration of Environmental Product type "Cradle to Gate" [52], the procedure of [53] was followed. The consumption of blue water in the life cycle of construction products was considered, it was done in 2 stages: Product, with 3 mandatory modules (A1, A2 and A3) and Construction, with 2 optional modules (A4 and A5), which are indicated:

Module A1 (Raw materials): extraction and processing of materials and production.

Module A2 (Transport): round trip distance from materials to concrete plant, fuel and efficiency.

Module A3 (Fabrication): manufacture of concrete and washing of mixing equipment.

Module A4 (Transport): round trip distance from the plant to the construction site, fuel and its efficiency.

Module A5 (Construction): cleaning work area, washing machines, curing concrete.

2.2.3. Carbon footprint. The type "Business-to-consumer" [54] was considered, with the principles of [55], [56] and the procedure of [57]. The equivalent carbon emissions in the life cycle of a product were considered, developing in 5 stages that are indicated:

Raw Material: production of aggregates, cement and superplasticizer additive.

Transport: transportation of materials to the concrete plant according to round trip distances.

Production: concrete manufacturing process in the plant.

Distribution: transport of the prepared concrete to the project according the round trip distance.

Construction: machines used in placing concrete.

3. Results and analysis

3.1. Compressive strength

Figure 1 shows the effect of w/c on the compressive strength at 28 days, it is seen that when changing from PC-030 to PC-0.32 the resistance rises from 252.00 kg/cm^2 to 266.5 kg/cm^2, which indicates a 5.75% increase over PC-030. [58], [59] studied mixtures with PC for w/c of 0.30, 0.35 and 0.27, 0.30, finding an increase in resistance of 12.31% and 1.16% when going from a lower to a higher w/c. [60] indicates that this is because the mixture is more fluid and better wraps the aggregate.

Table 1. Mix design for 1 m^3 of PC.

MATERIALS	PC-0.30	PC-0.32
Cement (kg)	433.44	419.64
Water (l)	130.03	134.28
Fine aggre. (kg)	1472.02	1472.02
Coarse aggre. (kg)	78.64	78.64
Superplasticizer (l)	4.71	4.57
Content void	13 (%)	13 (%)

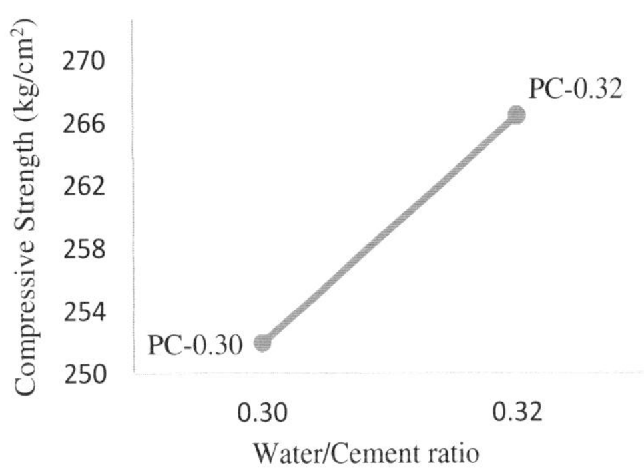

Figure 1. Effect of w/c on compressive strength.

3.2. Flexural strength

Figure 2 shows the effect of w/c on flexural strength at 28 days, it is seen that when changing from PC-030 to PC-0.32 the resistance increases from 34.10 kg/cm^2 to 37.40 kg/cm^2, which indicates an increase of 9.68% compared to PC-030. [58], [61] studied mixtures with PC for w/c 0.30, 0.35 and 0.34, 0.38 finding an increase in resistance of 4.20% and 20.01% when going from a lower to a higher w/c. [61] indicates that this is due to the greater hydration of the cement.

3.3. Permeability coefficient

Figure 3 shows the effect of w/c on the permeability coefficient at 28 days, it is seen that when changing from PC-030 to PC-0.32 the coefficient rises from 0.88 cm/s to 0.92 cm/s, which indicates a 4.55% increase compared to PC-030. [62], [63] studied mixtures with PC for w/c 0.28, 0.31 and 0.30, 0.35 finding an increase in resistance in 212.50% and 2.85% when going from a lower to a higher w/c. [61] indicates that this is due to decreased adhesion between set and aggregate cement.

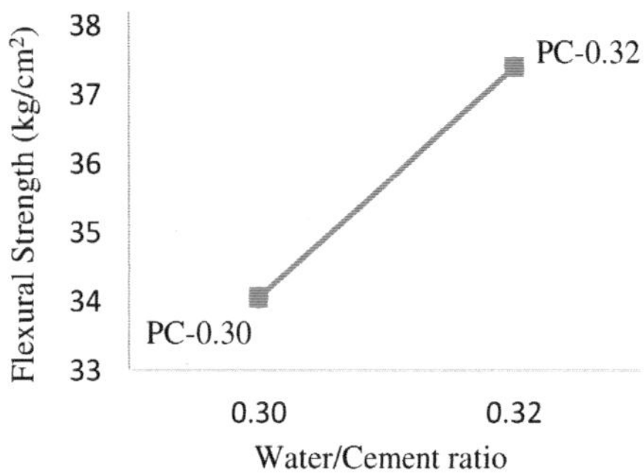

Figure 2. Effect of w/c on flexural strength.

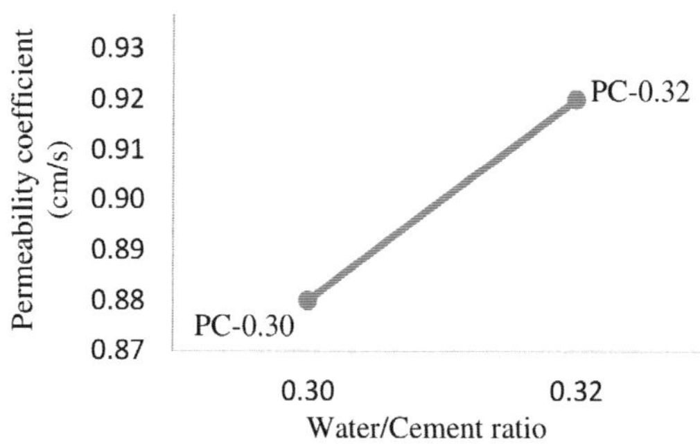

Figure 3. Effect of w/c on the permeability coefficient.

3.4. Surface runoff

Table 2 shows the surface runoff of a CC and a PC, it is seen that the runoff in the PC decreases by 42.43% after 30 minutes. [64], [65] evaluated the runoff of a permeable and a porous pavement using SWMM 5.0 software, and also compared several conventional development systems with a low impact development, finding a runoff reduction of 54.79 % and 98.3% in permeable pavement and in low-impact development. [66] indicates that this is because the precipitation falling on the permeable pavement surface infiltrates into a deposit that stores it.

Table 2. Surface runoff from CC and PC.

	Runoff		
Hours	**CC (l/s)**	**PC (l/s)**	**Reduction (%)**
00:30:00	2074.30	1194.17	42.43
01:00:00	2642.78	1194.17	54.81
01:30:00	1805.78	161.25	91.07
02:00:00	1703.55	161.25	90.53
02:30:00	1429.34	0.00	100.00
03:00:00	1386.33	0.00	100.00

3.5 CC and PC unit prices

Table 3 and Table 4 show the analysis of unit prices per m^3 of CC and PC, it is seen that the unit price of S/357.43 ($ 99.87 USD) for PC is 1.15% less than S/361.57 ($ 101.03 USD) of CC. [67] compares the costs of the pavements of 2 parking lots 0.15m thick built with CC and PC, finding that the cost of the PC pavement is 6.19% less. [68] compares the cost of a conventional development technique with that of permeable pavement in a project, finding that the cost of permeable pavement is 19.64% lower. [69] indicates that the cost of the PC is lower because it minimizes the need to use runoff retainers.

Table 3. Analysis of unit prices of 1m^3 of CC.

Description	Unit	Crew	Quantity	Price S/.	Partial S/.	Total S/.
Materials						**361.57**
Water	m^3		0.2300	5.68	1.31	
Fine sand	m^3		0.5775	42.29	24.42	
Portland cement	bag		10.8000	21.60	233.28	
Crushed stone 1/2"	m^3		0.3853	51.61	19.89	**278.89**
Workforce						
Foreman	hh	0.2000	0.0727	23.08	1.68	
Operator	hh	2.0000	0.7273	19.23	13.99	
Official	hh	2.0000	0.7273	15.94	11.59	
Pawn	hh	8.0000	2.9091	14.33	41.69	**68.94**
Teams						
Manual tools	%WF		3.0000	68.94	2.07	
Concrete mixer	hm	1.0000	0.3636	26.60	9.67	
Concrete vibrator	hm	1.0000	0.3636	5.48	1.99	**13.73**

Table 4. Analysis of unit prices of 1m^3 of PC.

Description	Unit	Crew	Quantity	Price S/.	Partial S/.	Total S/.
Materials						**357.43**
Water	m^3		0.1343	5.68	0.76	
Fine sand	m^3		0.0300	42.29	1.27	
Coarse aggregate ¾"	m^3		0.5600	51.61	28.90	
Portland cement	bag		9.8739	21.60	213.28	
Superplasticizer 971	lt		4.5700	7.12	32.54	**276.75**
Workforce						
Foreman	hh	0.2000	0.0727	23.08	1.68	
Operator	hh	2.0000	0.7273	19.23	13.99	
Official	hh	2.0000	0.7273	15.94	11.59	
Pawn	hh	8.0000	2.9091	14.33	41.69	**68.94**
Teams						
Manual tools	%MO		3.0000	68.94	2.07	
Concrete mixer	hm	1.0000	0.3636	26.60	9.67	**11.74**

3.6. Water Footprint of CC and PC

For the calculation of the Water Footprint (in l) of Conventional Concrete (WF_{CC}) and of the Water Footprint (in l) of Porous Concrete (WF_{PC}), equation (1) was applied:

$$WF_{CC} = WF_{PC} = WF_{A1} + WF_{A2} + WF_{A3} + WF_{A4} + WF_{A5} \quad (1)$$

where, WF_{A1}: Water footprint of module A1 (l), WF_{A2}: Water footprint of module A2 (l), WF_{A3}: Water footprint of module A3 (l), WF_{A4}: Water footprint of module A4 (l), WF_{A5}: Water footprint of module A5 (l). We find WF_{A1} with (2):

$$WF_{A1} = WF_{FA,CA} + WF_{CE} + WF_{SA} \quad (2)$$

where; WF_{FA},WF_{CA},WF_{CE},WF_{SA}: Water footprint of fine and coarse aggregates (0.73 l) [70], cement (0.73 l) [70], superplasticizer additive (0.73 l) [70]. Replacing in (2): WF_{A1}= 159.93 l (a) and WF_{A1}= 162.37 l (b). We find WF_{A2} with (3):

$$WF_{A2}= 2 \, x \, [DC \, x \, DW \, x \, WF_D \, (D_{FA} + D_{CA} + D_{CE} + D_{SA})] \quad (3)$$

where; 2: number of round trips, DC: Diesel consumption (0.71 l/m^3) [70], DW: Diesel weight (0.85 Kg/l) [70], WF_D: Diesel water footprint (0.05 l/Kg) [70], D_{FA}, D_{CA}, D_{CE}, D_{SA}: Transport distances of the supplier of fine and coarse aggregates (35.80 Km) [44], cement (9.20 Km) [44], superplasticizer additive (7.4 Km) [44], to the ready mix concrete plant. Replacing in (3): WF_{A2}= 1.79 l (c) and WF_{A2}= 1.99 l (ch). We find WF_{A3} with (4):

$$WF_{A3} = WM + WC/AC \quad (4)$$

where; WM: Amount of water for mixing (230 l), (134.28 l) [51], [71], WC: Amount of water for cleaning mixing equipment (5000 l) [70], AC: Amount of concrete for a concrete mixer with a discharge capacity of 3m3 (6.80 tn = 6800 Kg) [70], (6.33 ton= 6330 Kg) [71]. For the PC, the WC value was obtained considering that this concrete uses a lower w/c ratio to guarantee a good adhesion of the cement paste with the aggregates [72], therefore the amount of water in the mix will be less than the CC. Having considered the amount of 4650 l, obtained by discounting 7% which is the difference between materials per m^3 of a CC and a PC. Replacing in (4): WFA3= 964.86 l (d), WF_{A3}= 968.88 l (e). We find WF_{A4} with (5):

$$WF_{A4} = 2 \, x \, (DC \, x \, DW \, x \, WF_D \, x \, D_{PO}) \quad (5)$$

where, 2: Constant that represents the number of round trips, DC, DW, WF_D: indicated when calculating WF_{A2}, D_{PO}: Distance from the ready-mix concrete plant to the work (23.30 km) [44]. Replacing in (5): WF_{A4} = 1.58 l (f). We find WF_{A5} with (6):

$$WFA_5 = K \, x \, (WM/AC) \quad (6)$$

where; WM, AC: indicated when calculating WF_{A3}, K: Variable that depends on the type of concrete (K = 0.06 conventional and K = 0.02 porous). The K values were defined based on the greater loss of moisture from an air curing compared to a curing with a protective blanket, being approximately 3 times more [73], having considered a K value = 0.06 for the case of CC. Replacing in (6): WF_{A4}= 344.12 l (g), WF_{A4}= 107.21 l (h). Replacing (a), (c), (d), (f), (h) and (b), (ch), (e), (f), (h) in (1): WFCC= 1472.28 l, WFPC= 1142.03 l.

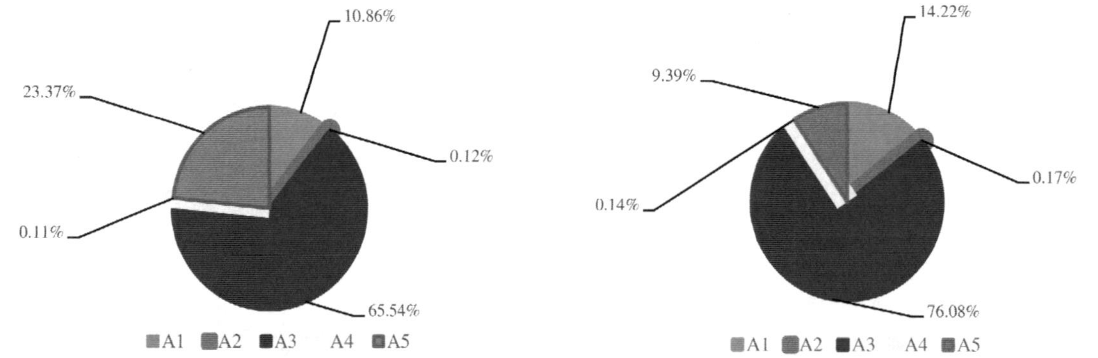

(a) (b)

Figure 4. Water footprint of CC (a) and PC (b) referenced by percentages of their modules.

Figure 4 presents the WF_{CC} and WF_{PC}, appreciating that the A3 module has a greater incidence in both specimens, reaching a higher value for the PC of 76.08%, which represents 16.08% more compared to 65.54% of the CC. Likewise, we must indicate that the water footprint for the CC is 1472.28 l and 1142.03 l for the PC, which represents 22.43% less; This difference is associated with Module A5 (Construction), which is obtained 344.12 l for the CC and 107.21 l for the PC, which is equivalent to 31.13% and represents 68.85% less, which is related to the lower amount of water Required for curing by using wet blankets versus CC pressure water jet curing. Additionally, we can mention that the difference is also related to Module A3 (Manufacturing) where it reaches 964.86 l for CC and 868.88 l for PC, which is equivalent to 90.05% and represents 9.95% less; being related to the least amount of water in the mixture and the washing of the plant. In this regard, some authors indicate that the greater water footprint of the CC is associated with the volume of concrete [74] and the waste of water [75].

3.7. Carbon footprint of CC and PC

The calculation of the Carbon Footprint (in $KgCO_2e$) of Conventional Concrete (CF_{CC}) and of the Carbon Footprint (in $KgCO_2e$) of Porous Concrete (CF_{PC}) was done with equation (16):

$$CFCC=CFPC = CF_{RM}+ CF_{TR} + CF_{PD} + CF_{DI} + CF_{CO} \tag{7}$$

where, CF_{RM}: Carbon Footprint of the Raw Material for the materials ($KgCO_2e$), CF_{TR}: Carbon Footprint of the Transport from the supplier to the concrete plant ($KgCO_2e$), CF_{CD}: Carbon Footprint of the Production in the concrete plant ($KgCO_2e$), CF_{DI}: Carbon footprint of the Distribution using the transport from the concrete plant to the work ($KgCO_2e$), CF_{CO}: Carbon footprint of the Construction using different machines for placing the concrete ($KgCO_2e$). We find CF_{RM} with (8):

$$CF_{RM}= CF_{FA}+ CF_{CA}+ CF_{CE}+ CF_{SA} \tag{8}$$

where, CF_{FA}, CF_{CA}: Carbon footprint of fine and coarse aggregates, CF_{CE}: Carbon footprint of cement, CF_{SA}: Carbon footprint of superplasticizer additive. The variables CF_{FA} and CF_{CA} are found with (9) and (10):

$$CF_{FA}= 4.7 \times FA/1000 \tag{9}$$
$$CF_{CA}= 4.7 \times CA/1000 \tag{10}$$

where, 4.7: Constant for fine and coarse aggregates ($KgCO_2e$ /ton) [76], 1000: Conversion factor (Kg/ton), FA, CA: Quantities of fine aggregate (924 Kg) [51], (78.64 Kg) [71] and coarse aggregate (655 Kg) [51], (1472.02 Kg) [71]. Replacing in (9) and (10): CF_{FA} = 4.34 $KgCO_2e$ (i), CF_{CA} = 3.08 $KgCO_2e$ (j); CF_{FA} = 0.37 $KgCO_2e$ (k), CF_{CA} = 6.92 $KgCO_2e$ (l). The variable CF_{CE} is obtained with (11):

$$CF_{CE}= 800 \times N/1000 \tag{11}$$

where, 800: Constant for the cement ($KgCO_2e$/ton) [57], 1000: Conversion factor (Kg/ton), N: Quantity of cement (459 Kg) [51], (419.64 Kg) [71]. Replacing in (11): CF_{CE}= 367.20 $KgCO_2e$/ton (m), CF_{CE}= 335.71 $KgCO_2e$/ton (n). The variable CF_{SA} is obtained with (12):

$$CF_{SA}= 1.17 \times AS \tag{12}$$

where, 1.17: Constant for the superplasticizer additive ($KgCO_2e$/Kg) [77], SA: Amount of the superplasticizer additive (4.57 Kg) [71]. Replacing in (12): CF_{SA}= 5.35 $KgCO_2e$ (ñ). Replacing (i), (j), (ñ) and (k), (l), (n), (ñ) in (8): CF_{RM}= 374.62 $KgCO_2e$ (o), CF_{RM} = 348.35 $KgCO_2e$ (p). We find HC_{TR} with (13):

$$CF_{TR}= 2 \times F\ (D_{FA} + CF_{FA} + CF_{CE}+ CF_{SA}) \tag{13}$$

where, 2: number of round trips, F: CO_2e emission factor per km traveled (2) $kgCO_2e/km$) [78], D_{FA}, D_{CA}, D_{CE}, D_{SA}: transport distances of the supplier of fine aggregates (35.80 Km) [44], coarse (35.80 Km) [44], cement (9.20 Km) and superplasticizer additive (7.40 Km) [44] to the ready mix concrete plant (Km). Replacing in (13): CF_{TR} = 323.20 $kgCO_2e$ (q), CF_{TR} = 352.80 $kgCO_2e$ (r). We find, CF_{PD} = 271.00 $kgCO_2e$ considering 271.00 $kgCO_2e$ (emission in concrete plant) [79] (s). We find HC_{DI} with (14):

$$CF_{DI}= 2 \ x \ F \ x \ D_{RC} \tag{14}$$

where, 2: number of round trips, F: considered in CF_{DI}, D_{RC}: Transport distance from the ready mix concrete plant to the work (26.30 km) [44]. Replacing in (14): CF_{DI} = 105.20 $kgCO_2e$ (t). We find CF_{CO} with (15), (16), (17). We find CF_{CO} with (15), (16), (17):

$$CF_{CO}= E_{MC} + E_{CC} + E_{VC} \tag{15}$$

$$E_{MC} = E_{CC} = E_{VC}= CH_{MC,\ CC,\ VC} \ x \ HT_{MC,\ CC,\ VC} \ x \ EF_{\ MC,\ CC,\ VC} \tag{16}$$

$$E_{MC}= E_{CO} = HC_{MC,\ CO} \ x \ WH_{\ MC,\ CO} \tag{17}$$

where, CF_{CO}: Carbon footprint of the machinery used in placing the concrete ($kgCO_2e$), $CF_{MC,CC,VC,CO}$: Hourly fuel consumption of mixer (40 l/h) [57], cutter (51 l/h) [57], vibrator (11 l/h) [57], and compactor (16 l/h) [57], $WH_{MC,CC,VC,CO:}$ Concrete mixer working hours (4.62 h) [57], cutter (0.37 h) [57], vibrator (289.74 h) [57], and compactor (h), $EF_{MC,CC,VC,CO}$: Emission factor (2.47 Kg CO_2e / l) [57] for concrete mixer, cutter, vibrator and compactor. Replacing in (16) for CC: E_{MC} = 456.46 $kgCO_2e$ (u), E_{CC} = 46.61 $kgCO_2e$ (v), E_{VC} = 7872.24 $kgCO_2e$ (w) and with these in (15): CF_{CO} = 8375.31 Kg CO_2e (x). Replacing in (17) for PC: E_{MC} = 456.46 $kgCO_2e$ (u), E_{CO} = 7345.98 $kgCO_2e$ (y) and with these in (15): CF_{CO} = 7802.44 Kg CO_2e (z). Replacing (o), (q), (s), (t), (x) and (p), (r), (s), (t), (z), in (7): CF_{CC} = 9449.33 Kg CO_2e, CF_{PC} = 8879.79 $KgCO_2e$.

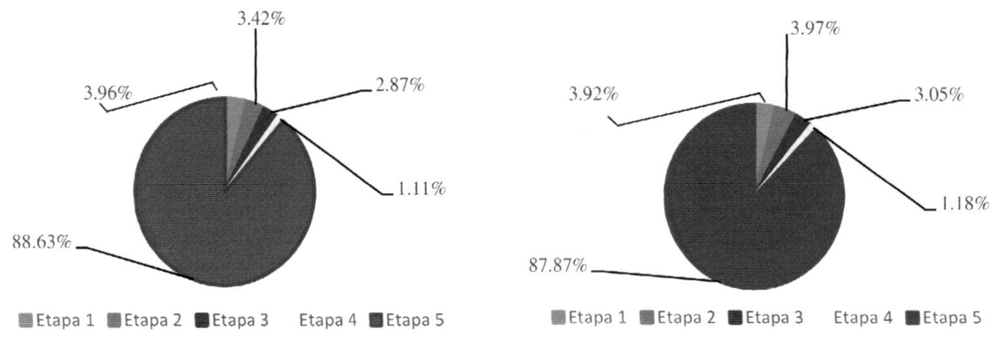

Figure 5. Carbon footprint of CC (a) and PC (b) referenced by percentages of their stages.

Figure 5 presents the CF_{CC} and CF_{PC} referenced by stages, where it is seen that stage 5 Construction has a greater incidence in both concrete, reaching a higher value for the CC of 88.63%, which represents a 0.86% more compared to 87.87% of the PC. Likewise, we can indicate that the carbon footprint for the CC is 9449.33 $kgCO_2e$ and 8879.79 $kgCO_2e$ for the PC, which represents 6.03% less, which is related to the use of the vibrator and concrete cutter. Additionally, we can mention that the difference is also related to stage 1 Raw Material, where it reaches 374.62 $kgCO_2e$ for CC and 348.35 $kgCO_2e$ for PC, which is equivalent to 92.99% and represents 7.01% more; being related to the greater amount of cement required for the preparation of concrete. On the other hand, it is worth mentioning that in stage 2 Transport, the PC is 352.80 $kgCO_2e$ and the CC 323.20 $kgCO_2e$, being higher by 9.16%, due to the use of the superplasticizer additive. In this regard, some authors indicate that PC has greater reflectivity in pavements [80], produces an increase in pavement albedo that reduces CO_2 emissions [81] and acts as a sustainable strategy to reduce CO_2 emissions [82].

4. Conclusions

The lack of a drainage system in cities affected by El Niño causes flooding and the collapse of the sewerage system, altering the quality of life in regions and countries.

The higher w/c ratio allows sufficient hydration of the cement paste and leads to increased compressive strength of the concrete.

The increase in flexural strength is due to a better bond between the cement and the aggregates.

The increasing permeability coefficient contributes to a greater collection of rainwater due to the non-clogging of the concrete pores.

The reduction in surface runoff is due to the amount of its voids, which guarantees the effective circulation of water and the adequate management of rainwater.

The lower cost of porous concrete is due to its lower quantity of component materials compared to those used in conventional concrete.

The lower water footprint in a porous concrete is related to the infiltration of surface runoff waters.

The lower carbon footprint of porous concrete reduces carbon emissions and lowers the heat island effect.

5. References

[1] Yan J 2020. *Temporal Convolutional Networks for Prediction of ENSO. Sc Rep*, 10, p. 1-15.
[2] Cornejo M 2017. *Gestión de riesgos: El caso de El Niño / Oscilación sur* p.1-72.
[3] Trenberth K 1997. *The Definition of El Niño. Nat Center for Atm Res*, 78, p. 2771-2777.
[4] Aziz O. *El Niño: Predicted hydrologic response in the US. Env &Water Res 2010*, 9 p.
[5] Glantz M. *Once Burned: Lessons Learned from 1997-98 El Niño, Un Press*, 2001. p. 1-313.
[6] Rocha A 2017. *El meganiño 1997-98*. Lima, p. 1-53.
[7] CEPAL 1998. *Ecuador: Evaluación efectos socio-económicos El Niño 1997-1998*, p. 1-78.
[8] Tucci C 2007. *Gestión de Inundaciones Urbanas. World Met Org*, 315, p. 1-317.
[9] Karabegović I 2019. *New Technologies, Development and Application II*. p. 790-797.
[10] Bai Y. *Evaluation System for Optimal Allocation of LID Facilities. Water* 2019, 11, p. 341.
[11] Shrivastava A 2018. *Global Warming for Sustainable Drainage System, Co Tech*. p. 15-28.
[12] Singh R. *Investigación experimental sobre hormigón permeable, Civ. In*. 2020, p. 229-240.
[13] Barišić I 2017. *Pervious concrete mix for sustainable pavement solution. Earth Env*, p. 1-7.
[14] Yahia A 2014. *New approach to proportion pervious concrete. Con Mat*, p. 38-46.
[15] ACI. *Report on pervious concrete. ACI-522, 2010*. p. 1–38.
[16] Hesami S 2014. *Effects of rice husk ash, fiber of pervious concrete. Con Ma*, p. 680-691.
[17] García E 2011. *Control de escorrentías urbanas mediante pavimentos permeables*, p 1-10.
[18] Mayz C 2018. *Diseño alternativo para implementar SUDS con software, Jo In*, p. 285-299.
[19] Wong T 2002. *Model for Urban Stormwater Improvement Conceptualization. Urb Dr*, 4 p.
[20] Ashley R 2016. *Using the multiple benefits of SuDS tool to deliver long-term benefits*, 4p.
[21] Lai Fu 2010. *Current Capabilities and Planned Enhancements of Sustain*, p. 3271-3280.
[22] Gironás J 2010). *A new applications manual for the Storm Management. Env Mo So*, 2 p.
[23] Zhuan Y 2018. *Experimental Research on Recycled Concrete Road Base. Co En Ge*, p. 8.
[24] Lian C 2010. *Optimum mix design of enhanced permeable concrete. Co Bu Ma*, 8 p.
[25] Liu W 2019. *Performance of New Permeable Concrete Materials, Nat Env Pol Tec*, 7 p.
[26] Borhan T 2020. *Experimental investigations on polymer pervious concrete. Con Mat*. 12.
[27] Tabatabaeian M 2019. *An innovative performance pervious concrete with epoxy. Bu Ma*, 2.
[28] Šešlija M. *Possibilities of pervious concrete in road construction. Teh Vjesn*. 2018, 13 p.
[29] Tennis P. *Pervious concrete pavements. EB302.02. PCA & NRMCA*, 2004, p. 1-32.
[30] Obla K 2010. *Pervious concrete-An overview. Indian Concrete Journal*, 84, p. 9–18.
[31] Zhou L. *Ecological impacts of green roofs and permeable pavements. J En Ma*.2018, 21p.

[32] Sotelo J, et al. *Huella hídrica, desarrollo y sostenibilidad en España*. 5, 2011.
[33] Cámara Valencia. *Cuaderno de Comercio y Sostenibilidad: Huella Hídrica*. 2017, p. 1-30.
[34] Larsen L 2015. *Urban climate & adaptation strategies. Ecol & Env*, 13, p. 486-492.
[35] Espíndola C 2012. *Huella del Carbono. Conceptos, Métodos de Estimación. Inf Tec*, 14 p.
[36] Goméz D 2017. *Evaluación geológica de las zonas afectadas El Niño 2017-Tumbes*, 71 p.
[37] Bayer A 2014. *An unforgettable event: Study 1997-98 El Niño in northern Peru. Dis*, 21 p.
[38] Torres V 2015. *A 10 años de privatización EMFATUMBES ¿Éxito o fracaso?* pp. 1-32.
[39] ASTM. *ASTM C1157, Performance Specification for Hydraulic Cement.*
[40] ASTM. *ASTM C33, Specification of Aggregates for Concrete.*
[41] ASTM *ASTM C494, Specification for Chemical Admixtures for Concrete.*
[42] ASTM. *ASTM C1602, Specification for Mixing Water Used in the Production of concrete.*
[43] USEPA 2015. Storm Water Management Model - User's Manual Version 5.1, pp.1–353.
[44] Google Maps. Recuperado 2020 de: https://www.google.com.co/maps/place/Lima.
[45] ASTM. *ASTM C192, Practice for Making and Curing Concrete Test Specimens.*
[46] ASTM. *ASTMC39, Method for Compressive Strength of Cylindrical Concrete Specimens.*
[47] ASTM. *ASTM C78, Method for Determination of Flexural Strength of Concrete.*
[48] ICONTEC. NTC 4483. *Método para determinar la permeabilidad del hormigón al agua.*
[49] Ureta L. *Revista Costos, Edición de Aniversario*. Abril, 2017, Lima, p. 1-108.
[50] Cámara Peruana Construcción. (2014). *Costos y Presupuestos en Edificaciones*, pp 1-375.
[51] Castillo, R. (2013). *Tabla de dosificaciones y equivalencias. UNACEM*. pp. 1-9.
[52] SIS. SS-EN 15804, *Sustainability of construction works, environmental product*, 11 p.
[53] Netz J, Sundin J 2015. *Water Footprint of Concrete*, pp. 1-72.
[54] BSI, *PAS 2050 Specification for life cycle grenhouse gas emission of goods, services*, 46 p.
[55] ISO. ISO-14067. *Greenhouse gases, Carbon footprint products, Requirements,* guidelines
[56] ISO. ISO-14064. *Part 1: Specification with guidance at the organization level.*
[57] Ecenarro L, Díaz T 2014. *Estudio y desarrollo de huella de carbono: vía ferroviaria*, 59 p.
[58] Liu H 2018. *Strength, permeability of pervious concrete with others porosities, Ap Sc*, 17 p.
[59] Mulyono T 2019. *Pervious Concrete for Permeable Pavement. Con Ear & Env Sci*, 16 p.
[60] Zhu H 2020. *Study Permeability, Recycled Aggregate Pervious Concrete, Fibers. Ma,* 19 p.
[61] Guo P, et al. (2011). *Pavement Performance of Steel Pervious Concrete*. ICTE, p. 1-6.
[62] Yang, H. 2018. *Experimental Study on Permeability of Concrete. Con Ear & Eng Sci*, 8 p.
[63] Ibrahim, 2020. *Hydraulic & strength characteristics of pervious concrete, Co Bu Ma*, 13 p.
[64] Zhu H 2019. *Simulation study of permeable pavement on reducin flood risk. Tr Sc Te*, 10 p.
[65] Wilson C 2015. *Comparison of Runoff Quality and Quantity, J En En*, 11 p.
[66] Fassman E 2010. *Urban Runoff Mitigation by Permeable Pavement System. J Hy En*, 11 p.
[67] Rehan T 2018. *Life Cycle Cost Analysis for Traditional, Permeable Pavement. Co Re*, 11 p.
[68] Weitman D 2008. *Reducing Stormwater Costs through LID Strategies and Practices*, p. 11.
[69] Parasivamurthy P 2010. *Improving Ground Water Recharge Using Pervious Concrete*, 9 p.
[70] Netz, J 2015. *Water Footprint Concrete. School of Architecture and Built Env*, 73 p.
[71] Aguirre B 2020. *Diseño óptimo de pavimento drenante de concreto con SWMM 5.1*, 25 p.
[72] Md. Safiudddin 2007. *Effect of different curing methods on properties concrete. Au J*, 9 p.
[73] Neville A., Brooks J., *Concrete Technology, Longman Group*, UK, 1997, p. 1-460
[74] Rajaei M 2014. *Case of Wind Plant Life Cycle Energy, Emissions,Water Footprint*, 16 p.
[75] Hoekstra A 2011. *The water footprint assessment manual, Earthscan*, p. 1-228.
[76] Garcia M 2016. *Análisis de huella de carbono de industria de concreto y agregados*, p. 145
[77] Albornoz J 2015. *Cálculo de huella de Carbono, hormigones geopoliméricos, Chile*, 106 p.
[78] BSI. 2008. *Guide to PAS 2050: How to assess the carbon footprint of good, services,* 59 p.
[79] HOLCIM. 2014. *EPD of Ready-Mix Concrete*, p.1-13.
[80] Lee K 2010. *Cool Pavements as Sustainable Approach to Green Streets & Highways*, 9 p.
[81] Akbari 2007. *Global cooling: Effect of urban albedo on global temperature. Conf, Bu* 17 p.
[82] Debnath B 2018. *Pervious concrete as alternative pavement strategy, J Pav Eng*, 17 p.

AUTHOR INDEX

Institute of Physics
Dirac House, Temple Back
Bristol BS1 6BE UK

ISSN: 1757-8981
ISBN 978-1-7138-3337-6

CURRAN ASSOCIATES INC.
proceedings
.com